the

Green Devotional

the Green Devotional

Active Prayers *for a* Healthy Planet

Karen Speerstra

Conari Press

First published in 2010 by Conari Press,
an imprint of Red Wheel/Weiser, LLC
With offices at:
500 Third Street, Suite 230
San Francisco, CA 94107
www.redwheelweiser.com

ISBN: 978-1-57324-459-6
Library of Congress Cataloging-in-Publication Data is available upon request.

Cover and text design by Donna Linden
Typeset in Priori Sans and Warnock
Cover photograph © VisionsofAmerica/Joe Sohm/Getty Images

Printed in Canada
TCP
10 9 8 7 6 5 4 3 2 1

Printed on 100% recycled paper.

For Josephine Elizabeth Speerstra . . .

and for every grandchild.

Contents

I am the breeze that nurtures all things green.
I encourage blossoms to flourish with ripening fruits. I
am the rain coming from the dew that causes
the grasses to laugh with the joy of life.

—Hildegard of Bingen

Acknowledgments

For this book's inspiration, all the credit goes to Caroline Pincus and her colleagues at Conari/Red Wheel/Weiser. I only leapt at the chance to help form it, invite in the Green Planetary Voices, and add my own voice to the mix. Many books have been written about our planet in crisis, but it is our expectation that this one, anchored in so much wisdom from many people, times, and places, will offer not only further clarity and inspiration, but even more important, hope.

Joyful thanks go to my "kitchen cabinet," who offered title suggestions, focus, and all around "Sophia" encouragement: Laura Baring-Gould, Sharon Bauer, Julia Blackbourn, Lori Borden, Ellen Donaldson Allen, Maren Donaldson Lewis, Kelly-Ann Falkenburg, Jan Field, Carol Frenier, Ellen Frost, Mary Ann Garrity, Ann Gilman, Penny Hauser, Jean Peterson, Sherry Rhynard, Diane Root, and Judy Walke. And a special thank-you to Genevieve Kirchman, who spent hours on my office sofa hearing me, steering me, and helping me see larger pictures.

The research around gathering these planetary voices was facilitated to a great extent by two other supportive friends: Kathy Hartman and her extensive environmental library at the Vermont Law School and Tim Wolfe who introduced me to the Dartmouth Library in Hanover, New Hampshire. Several local libraries stood ready with their own supportive librarians—where would we be without them? Amy, Lynne, Rhonda—you know who you are.

I cannot begin to thank all the authors represented here, but one in particular stands out. He's been valiantly working, teaching, organizing, and, most important, writing about all aspects of ecology and climate change for twenty years: Bill McKibben. The importance of the number 350 is carried by many, but like a carefully crafted bell, Bill's voice continues to ring out that number, sounding across our entire planet, calling us to action. It is my hope you will find all of these authors' books, in their entirety, and take them to heart. My special thanks to Hildegard of Bingen, whose voice starts us off.

To my family—particularly Josie, to whom this book is dedicated and who, before too long, will be able to read it all by herself—thank you for offering your encouragement and sometimes other nitty-gritty stuff whenever I ask for it: John, Joel, Nathan, Traci. My true loves.

When one tackles something as audacious as a "devotional," the unseen ones support you—particularly Sophia, the Divine Feminine Wisdom, also known as Gaia, who has upheld our earth from its very beginning. She continues, along with all our other guardians, to smile and wink as we feel our way. For surely we are not in this alone.

You will find more about Sophia at my web site: *www.sophiaserve.com.*

Finally, my thanks go to you. You're reading this. *You're the one* to make a difference.

Introduction

For some reason, perhaps known only to the archangels, we humans were given the gift of Choice. Free Will. It's hard-wired into us at birth. When we overlook or undervalue our choices, it's easy to imagine we're victims. Our mouths are quick to form questions like "Why me?" "Why this?" "Why now?"

In times of crisis—and planet earth IS in crisis—we look around for something to hang on to. And when we grab for it, we notice other people reaching out to us. We hear voices asking, "How can I help?" "What do you need?"

No time was this more evident than when New York's World Trade Center towers crumpled on September 11, 2001, spewing ash and fear of terrorism around the world. But across the street from the towers, a tiny Episcopal church called St. Paul's Chapel and its 200-year-old cemetery remained unscathed. Some say that this building, which resembles London's St. Martin-in-the-Fields, was spared because a large sycamore tree bore the brunt of the winds. Others claim that homeless people who slept there the night before had left some windows open, inadvertently keeping the little church from exploding. Whatever the reason, that morning's deadly breezes blew through one of Manhattan's oldest buildings, over George Washington's chair, past the handmade door hinges, grazing a painting of the Great Seal of the United States and—whoosh!—out the other side.

Overnight, St. Paul's Chapel became a haven for rescue workers—a place where they could sit, eat a snack, cry, sip some water, and maybe stretch out on a pew. It truly transformed into what a church is meant to be: a sanctuary. For eight months, volunteers came from all over to work twelve-hour shifts, to counsel and care for the firefighters, police, construction workers, and countless others needing to escape the aura of death, if only for a few quiet moments. Volunteer podiatrists set up shop in George Washington's boxed pew to heal bruised and battered feet. Who better to recognize the need to attend to injured feet than the General? Massage therapists, chiropractors, musicians, food servers—they all came to help. These volunteers embodied active prayer. Soon after the attack, the folks running St. Paul's decided to remove its historic pews, except for George Washington's and the pew where New York's first governor sat. As you might imagine, this decision was not without its conflicting viewpoints. But finally the church council agreed to take them out in order to accommodate all the makeshift cots, allowing weary workers to rest more easily. After the crisis was over, they decided to leave the pews out—all but one. Gouged by firefighter's equipment and scuffed by rubber boots, that one lonely pew stands near St. Paul's front doors to remind visitors of the heroes who momentarily rested there. St. Paul's has more flexible seating now, and its doors remain open as gentler winds continue to blow through.

Veriditas

As we learned during our Crayola days, yellow and blue combine to form green. Chemistry class taught us that copper turns green when a thin surface patina of complicated chemistry combines water, oxygen, and carbon dioxide. If you warm iodine crystals, you get a yellow-green fluorescence; green comes from algae in water, hydrogen in ice. Silicates and oxides bonding. Chlorophyll in leaves. Nitrogen in diamonds. Chemists tell us that the formula for emerald green is $Cu(CH_3COO)_2As_2O_4$.

But most of us don't speak in letters and subscripts, and a small percentage of us (more men than women) have eyes that lack green-sensitive cones. Even if we're not colorblind, we may overlook the color green. As a result, we sometimes miss the iridescence on a hummingbird's wing, the green flash of a dragonfly, the jeweled turquoise in a peacock feather, or the rare flash of green in the setting sun. And we may not be fully aware of how much earth needs us to green up. Right now.

This collection of "grace notes," of "green plainsongs," seeks to value all shades of green. Just as green ripples from yellowy-chartreuse to deep jade, each of us views our planet on a spectrum. For some, earth is still a paradise. For others, it's a garden. A home. A gift. A way station to something better. A test pad or proving ground. And for far too many of us, it's a rich supply of "goodies" to use or to hoard.

As we "green up" together, we sense many shades of our communal spirituality blowing through. We ponder who the "I" is in Hildegard of Bingen's words that begin this book. "I am the breeze that nurtures all things green." It doesn't matter how we define that "I" or how we describe and hold our individual beliefs. The important thing is that the same breeze wafts over us all. We hear voices in concert, but our spirituality takes many forms; the active prayers within these pages may not always fit your own definition of prayer. In her book *God in All Worlds,* Lucinda Vardey puts it this way: "Spiritual truth requires deep commitment to the divine in ourselves and others and can lead to transcended states of awareness, of consciousness that unites the mind, the heart and the soul. Spirituality is about being open to different realities of existence, about being guided by our intuition, which is nothing less than the truth within us."

In an attempt to find some of those truths within, we include here 350 bits of devotional wisdom. Three hundred fifty is a planet-saving number. Scientists such as NASA's Jim Hanson have warned us for a few years that when our atmosphere holds up to 350 parts-per-million of carbon dioxide, our earth has a fever. We know what fevers can lead to. In our planet's case, polar ice melts faster and faster. Ocean temperatures rise. Coral reefs turn gray. We hovered around the 280 mark for a while, which was high, but not terribly dangerous. But the figure has continued to rise. Now we're approaching "stroke" time. By burning more and more fossil fuels (coal's a big contributor—and there may be no such thing as "clean" coal as far as CO_2 is concerned), at the time of this writing we've reached 385. This is very bad news, because we've already passed earth's safety threshold, and no one is sure if we can bring her fever down. Soon we will likely reach the point of no return, unless we act with great concerted effort. To do that, governments will have to cooperate.

Bill McKibben, prolific author, professor, activist, and the fellow Vermonter's voice you'll often find in this book, once quipped: "You're going to fix global warming by changing lightbulbs? Try changing your politicians instead. Screw in a new Congressperson."

As President Obama often points out, government cannot reverse global warming by itself. But it was our government that did not regulate, that let corporations get us into this mess. Paradoxically, it is business, encouraged by our government, that will have to figure out how to reduce carbon footprints through offsets, paybacks, barters, and other innovative measures. Scientists gave us nuclear energy, and it is scientists who now must figure out how to keep us safe. *Each of us* is also to blame, so each of us now needs to seriously think about how we live here and how much carbon dioxide we're individually pumping out. Snow machines, for instance, pump out more carbon dioxide than emission-controlled cars. Even though most of us don't think much about chemistry, *carbon dioxide* is one chemical term we all need to understand. And 350 is a number that Bill McKibben and many others think should be ingrained onto our collective psyche. Church bells should ring out 350 times to remind people how important three-five-oh is to our very survival.

For those of us who use books like this one as a daily devotional resource, we've added sixteen more prayers in the final chapter—one extra for a leap year—for a total of 366 prayers. There are no dates attached. Your green devotional year might begin on Earth Day. Or your birthday. Or on January first. How you use this book is totally up to you.

Arne Næss, the mountain-climbing Norwegian philosopher who gave us the term "deep ecology," also coined the term *ecosophy.* It melds *ecology,* the study of relationships, with *sophia,* the Greek word for wisdom. This, then, could be called a little book of ecosophy, because it contains a selection of today's wisdom about earth—secular and sacred, ancient and current. It speaks to numerous topics related to climate change and the care of our planet and everything that lives here. These voices, in concert, were selected and organized to inspire, to enrich, to support, and sometimes to challenge us as we face the future together.

Kermit the Frog told us, "It's not easy bein' green." He's right, of course. But Hildegard of Bingen, the Rhineland mystic, reminded us that *greening* may well be the very reason we're here—*Veriditas,* she called it.

A Green Book of Hours

This collection of voices is a "green book of devotional hours," reminiscent of the Books of Hours people held in their palms during medieval times. They were called "cathedrals in your hands," and reminded their readers that they were connected to something broader and wiser than themselves. Those gold calligraphic and lavishly illustrated books were often organized according to the Church's eight canonical hours, which many monastic communities still use to measure out prayers in, roughly, three-hour increments: Lauds, the dawn praise, followed by Prime, Terce, Sext, None, Vespers, Compline, and finally, at darkest midnight, Matins. Some believe that the early monks broke up their days this way because they were influenced by the Vikings, who measured their time by the eight tides: Midnight, Dawn, Morning, Vaporlessness, Noon, Rest, Evening, and Shadow.

This eight-part book is "our planet in your hands." As the voices began to harmonize together, they called out for their own canonical structure—one bounded by the ancient elements of *Earth, Air, Fire,* and *Water.* To these elements we add four less tangible ones: *Time, Space, Essence,* and as many Books of Hours end, *Closing Prayers.* These prayers don't just ready us for sleep; they rejuvenate us and prepare us for another day of passionate action. For the voices within these pages call you not just to prayer, but *to action.* They form an eight-part polyphonic chorus in support of our green planet. They're a "shout out," a call to "step up."

Open this book anywhere, anytime, and hear what the voices are saying. Choose one of the devotional thoughts to think about—alone or with friends, meditate on it, and *act upon it.* For when we are devoted to something, we cannot help but fervently live it.

Blessed be the precious and preserving air, by which
we are given life.
Blessed be the precious and preserving fire, by which
we are warmed.
Blessed be the precious and preserving water, by which
we are cleansed.
Blessed be the precious and preserving earth, by which
we are sustained.
—Caitlin and John Matthews, *Walkers Between the Worlds*

1

Earth

Earth. Matter. Mater. Matrix. Mother. Gaia, the deep-breasted one. She's a fragile placenta, as the Maori say, birthing us all. As a living, breathing ecosystem, her body is composed from the same elements as yours and mine. Her skeleton, stones. Her lungs, trees. Sometimes she shakes and sneezes ash. Over the centuries, she's gotten sick and then recovered, given enough time. But there are so many of us humans, now, she can hardly keep up. We've got to lower her fever before she suffers a stroke.

Like Goldilocks hopscotching through the woods, we convince ourselves that everything will be just fine if we try hard enough, develop the right science and technology, and raise enough money to eat the just-right porridge, relax in the just-right chair, and stretch out on the just-right bed. But our greedy consumption has upset the earth's balance, and the "bears" are coming back to see who's broken into their house.

Some people argue that this "climate change stuff" is all just another "normal cycle." Not to worry, they say. It's just inflated hype. Other voices claim we may have ten years of normalcy, at best. Others say we may not even have that long before our polar ice melts and our oceans begin to substantially change. How much real trouble is our planet in?

According to the World Health Organization, climate change has already directly or indirectly killed more than a million people globally since 2000. Most of those affected died from fevers for which we have no vaccine or cure, like the dengue fever, which has taken the lives of countless people living in Asia. Many more have perished from flooding and landslides. Coming dust storms and wildfires will lower crop yields and cause more salt buildups in lower-lying areas, affecting our food and water supplies.

Scientists have presented various models to show us how our planet will likely change in the coming centuries.

An article by Andrea Thompson and Ker Than, from the website *LiveScience.com,* cites statistics from the Intergovernmental Panel on Climate Change, among other credible sources, to give us an idea of what our world will look like in the next two hundred years. Sadly, many experts now believe this scenario is most likely a 90 percent *fait accompli:*

> 2008–2018: Our oil production peaks. Some say this will occur by 2020. This will trigger a global recession, food shortages, and wars over dwindling oil supplies.
>
> 2020: Flash floods in all parts of Europe; less rainfall in other parts of the world. World population will reach 8.3 billion people.
>
> 2030: Diarrhea-related diseases will likely increase. Eighteen percent of our coral reefs will disintegrate; up to 30 percent will disappear from Asian coastal waters. Glaciers in Africa will have disappeared. Cities' populations will increase by 20 percent.
>
> 2040: The Arctic Sea will be ice-free in the summer. (Other scientists say this won't happen until 2060–2105.)

2050: Small alpine glaciers will be gone. Large glaciers (most of them in Greenland and Antarctica) will shrink by 70 percent. World population reaches 9.4 billion people.

2070: All glaciers will have disappeared, and hydropower stations will no longer be effective energy sources. Europe will be hardest hit.

2080: Some parts of our planet will have floods; others, droughts. Coastal flooding will become an ordinary occurrence. Sea levels around New York City could rise three feet. Between 1.1 and 3.8 billion people will experience water shortages and up to 600 million will go hungry.

2100: Experts believe temperatures could increase two to eleven degrees Fahrenheit; sea levels could rise by two feet. A combination of global warming and other factors will push many ecosystems to their limits. Twenty to thirty percent of all earth species could become extinct. Thawing permafrost will emit methane and more carbon dioxide than our atmosphere can absorb. "Dust Bowls" similar to what Americans experienced in the 1930s will be common.

2200: Days will be slightly shorter; waters will shift toward the poles; more mass there will speed up the earth's rotation.

Many of us suspected our planet was in serious trouble back when Congress passed the Clean Air Act of 1963. Citizens fought over the construction of dams and the cutting down of trees. Soon there was an oil embargo, and we parked in long lines at the gas pumps. Jimmy Carter put solar panels on the White House and encouraged us to drive slower, don sweaters, and turn down our thermostats to save energy. Reagan removed the White House solar panels and before we realized it, we'd removed our sweaters and were once again speeding down our freeways. We lost sight of those twelve inalienable rights, as Paul Ehrlich stated them.

1. The right to eat well.
2. The right to drink pure water.
3. The right to breathe clean air.
4. The right to decent, uncrowded shelter.
5. The right to enjoy natural beauty.
6. The right to avoid regimentation.
7. The right to avoid pesticide poisoning.
8. The right to freedom from thermonuclear war.
9. The right to limit families.

10. The right to educate our children.
11. The right to have grandchildren.
12. The right to have great-grandchildren.

Listen to the earth. She's stretching and shifting. Listen to her thrumming deep down. Listen to her groan as her glaciers calve and as bits of blue ice float away, making burping sounds as the oxygen is released. Can you hear the snows running off exposed rock on Kilimanjaro and trickling over open alpine fields? Her voice is on the wind. Her natural rhythms pulse through the night. Her rhythms are, indeed changing. Ever so slightly.

Somewhere along the line, we lost our way. "Small is beautiful" morphed into "big and more is better." Thoreau's caution to "simplify, simplify," disappeared in our rush to shop, eat, drive, build, and generally enjoy ourselves. The result is a planet in financial and ecological crisis.

The Theosophists may have it right: sow a thought and reap an action; sow an action and reap a habit; sow a habit and reap a destiny. We have sown, and now we are reaping.

We humans love shopping for "things," and it's even easier when all we need to do is click a computer key. Now, we're reaping our destiny of consumerism.

William Irwin Thompson, the mythologist and historian, said that there are only three essential questions for us as humans: What are we?, Where do we come from?, and Where are we going? If we are honest with ourselves, the answer to the first question is likely to be "consumers." This pretty much answers question number three.

On average, financial meltdowns notwithstanding, Americans spend an average of six hours a week shopping and, by contrast, only forty minutes a week playing with our children. And here's a startling statistic: 93 percent of American girls report store-hopping as their favorite activity. We have more shopping centers than high schools in America. The Mall of America opened in 1992 and soon became America's premier tourist attraction—topping even Disney World. The Grand Canyon attracts one tenth the number of people who visit Bloomington, Minnesota's favorite mall. This mega-complex has a chapel if you wish to be married, a university branch if you wish to be educated, and about seventy shops that purport to "sell nature." You can eat in a "rainforest café" to feed your fantasy of "living in the great outdoors" when all you're really doing is eating in an air-conditioned bubble-wrapped capsule. You can imagine you're out in nature without getting your hands dirty or being bothered by mosquitoes. Northwest Airlines maintains a fleet of shuttle buses to bring tourists directly to the mall from the nearby airport; that's all many of them see of our vast country. Malls are touted as public space, but they're privately owned and the only vote or voice you have there is your credit card.

We're beginning to understand the distinction between sustainable cultures that leave something behind for future generations and ravaging cultures that grab everything in sight. (Guess which one we belong to.) Daniel Quinn had no idea that his 1992 book about Ishmael, a talking gorilla, would become a publishing phenomenon. It took one person publishing one book (which is not only read casually, but has been adopted in high school and college courses on subjects a diverse as biology, sociology, psychology, world history, anthropology, philosophy, religion, political science, and economics, and literature) to reach millions of readers. Quinn's mentor, Ishmael the gorilla, encourages us each to teach one hundred people to be leavers, not takers, and to pass that message on exponentially.

Now we await the tipping point. Therein lies our hope, and Mother Earth's recovery.

Green Voices in Concert

The Mother of us all,
The oldest of all,
Hard, splendid as rock.
Whatever there is that is of the land
Is she who nourishes it . . .
—Homeric Hymn to Earth, 6500 BCE

That the conditions of life are violated, that the will of God does not triumph, that the beasts of the field are disorganized, that the birds of the air cry at night, that blight reaches the trees and the herbs, that destruction spreads among creeping things—this, alas! Is the fault of government.
—Chuang Tzu

If we surrendered
to earth's intelligence
we could rise up rooted, like trees.
—Rainer Maria Rilke

Deep bosomed Earth, sweet plains and fields fragrant
grasses in the nurturing rains,
around you fly the beauteous stars, eternal and divine,
come, Blessed Goddess, and hear the prayers of your
children,
and make the increase of the fruits and grains your
constant care,
with the fertile seasons your handmaidens,
draw near, and bless your supplicants.
—from Orphic Hymn to Gaia

My words are tied in one
with the great mountain
with the great rocks
with the great trees.
In one with my body and my heart.
Will you all help me,
with supernatural power,
and you, day
and you, night!
All of you see me
One with the Earth.
—California Yokuts prayer, quoted in *Gaia's Hidden Life*, Shirley Nicholson and Brenda Rosen

The way to love anything is to realize that it might be lost.
—G.K. Chesterton

We call upon the earth, our planet home, with its
beautiful depths and soaring heights,
its vitality and abundance of life, and together we ask
that it:
Teach us, and show us the way.
—Chinook blessing

Heart of Sky; Heart of Earth
give us our sign, our word,
as long as there is day, as long as there is light.
When it comes to the sowing, the dawning,
Will it be a greening road, a greening path!
—Matsuo Bashō

The Earth is the birthplace of our species and, so far as we know, our only home . . . . We are close to committing—many would argue we are already committing—what in religious language is sometimes called Crimes against Creation.

—from "Preserving and Cherishing the Earth, An Appeal for Joint Commitment in Science & Religion," signed by Carl Sagan, Freeman Dyson, Stephen Schneider, Peter Raven, Roger Revelle, and Stephen Jay Gould

God loves the earth fully. By loving one another and every sentient being—even the rocks who cry out—we love God. In this love we are called to resist the poisoning of peoples and the earth.

—Karen Baker-Fletcher, *Sisters of Dust, Sisters of Spirit*

Once you have lived on the land, been a partner with its moods, secrets, and seasons, you cannot leave. The living land remembers, touching you in unguarded moments, saying, "I am here. You are part of me."
—Ben Logan, *The Land Remembers*

There clings to the image [of Gaia] something of an older and once universal natural philosophy that quite spontaneously experienced the earth as a divine philosophy that quite spontaneously experienced the earth as a divine being animated by its own moods and intentions, the primordial Mother Earth.
—Theodore Roszak, *Person/Planet*

Brothers and Sisters: We cannot adequately express our feelings of horror and repulsion as we view the policies of industry and government in North America which threaten to destroy all life. Our forefathers predicted that the European Way of Life would bring a Spiritual imbalance to the world, that the Earth would grow old as a result of that imbalance. Now it is before all the world to see—that the life-producing forces are being reversed, and that the life-potential is leaving this land. Only a people whose minds are twisted beyond an ability to perceive truth could act in ways which will threaten the future generations of humanity.

—People of the Long House, Declaration of the Iroquois

I don't know what Nature is: I sing it.
I live on a hilltop
In a solitary whitewashed cabin.
And that's my definition.

—Fernando Pessoa

One might say that on the seventh day, when God rested, man took over as Creator. He began to redesign the world and mold it to his own liking. Century after century, he has worked without a blueprint to build a man-made world, each inventor throwing in an idea, each mason thrusting in a stone wherever a hand could reach. Only in very recent years have we begun to wonder what we are building.
—Thor Heyerdahl, speech, 1972

We must ask ourselves if we want to belong to the earth on which we walk or to be cut off forever from our roots, bereft of a sense of greater belonging.
—Caitlin and John Matthews, *Walkers Between the Worlds*

I am pessimistic about the human race because it is too ingenious for its own good. Our approach to nature is to beat it into submission. We would stand a better chance of survival if we accommodated ourselves to this planet and viewed it appreciatively instead of skeptically and dictatorially.

—E. B. White

Accomplishment of a program of integral survival of the planet, and of the human community, requires that the dominant profit motivation of the corporation endeavor be replaced with a dominant concern for the integral life community. To seek the benefit for humans by devastating the planet is not an acceptable project . . . We will change or we will die in a major part of our inner being.

—Thomas Berry, *The Great Work*

The sun flared down on the growing corn day after day until a line of brown spread along the edge of each green bayonet. The clouds appeared, and went away, and in a while they did not try any more. The weeds grew darker green to protect themselves, and they did not spread any more. The surface of the earth crusted, a thin hard crust, and as the sky became pale, so the earth became pale, pink in the red country and white in the gray country.

—John Steinbeck, *The Grapes of Wrath*

Our faith imposes on us a right and duty to throw ourselves into the things of the earth.

—Pierre Teilhard de Chardin, *Hymn of the Universe*

The great climatic cycles, biological evolution, and the natural changes of the landscape are all part of the earth's repertoire of balancing movements that keep it on its own tightrope. . . . The bad news is that mankind has so profoundly transformed the earth that nature's ability to produce the goods and services that are essential for human life . . . has become unpredictable. The good news is that mankind has given proof throughout its history of a great capacity to adapt to changes in the environment . . . [but] it will require major social and economic reforms over a period of several decades.

—Eric Lambin, *The Middle Path*

It is not like we're on the Titanic *and we have to avoid the iceberg. We've already hit the iceberg.*

—Rob Watson

So severe and so irreversible is this deterioration that we might well believe those who tell us that we have only a brief period in which to reverse the devastation that is settling over the Earth. Only recently has the deep pathos of the Earth situation begun to sink into our consciousness . . . such is the context in which we must view this transition period . . . as a moment of grace. A unique opportunity arises. . . . The story of the universe is now being told as the epic story of evolution by scientists . . . the one story includes us all. We are, everyone, cousins to one another. . . . Although the human challenges to these purposes must never be underestimated, it is difficult to believe that the larger purposes of the universe or of the planet Earth will ultimately be thwarted.

—Thomas Berry, *The Great Work*

If consumer society has one Achilles' heel, it's not that it is going to destroy the earth—it is, but that's not the Achilles' heel. The Achilles' heel is that consumer society doesn't make us unbelievably happy.
—Bill McKibben, quoted in *Visionaries: People and Ideas to Change Your Life*, Jay Walljasper and Jon Spayde

The Titanic *sank not because its captain did not see the iceberg that the ship was headed towards, but because he saw it too late.*
—Eric Lambin, *The Middle Path*

The major problems in the world are the result of the differences between the way nature works and the way people think.
—Gregory Bateson

We're like people living at the top of the world's tallest skyscraper who every day go down to the lower floors and knock two hundred bricks out the walls at random. We use these bricks to extend our living space, to build upward. Hey—two hundred bricks, that's nothing. There are millions in those walls down there. But every day the structural integrity of the building is being compromised—and there'll come a day when all these compromises connect up, and the whole thing will come down—not in a week or a day or even an hour. It'll come down all at once, in minutes.
—Daniel Quinn, Bioneers Conference, 2005

The way I like to put it is that we're driving in a car with bad brakes in a fog and heading for a cliff. We know for sure now that the cliff is out there, we just don't know exactly where it is. Prudence would suggest that we should start putting on the brakes.
—John Holdren

No one knows precisely how fast the coming changes will engulf us. If they come rapidly, they may spur us to action; if they come more slowly, we may dally.
—Bill McKibben, *Deep Economy*

To lovers of the wild, these mountains are not a hundred miles away. Their spiritual power and the goodness of the sky make them near, as a circle of friends.
—John Muir, *A Thousand-Mile Walk to the Gulf*

The soil is the great connector of lives, the source and destination of all. It is the healer and restorer and resurrector, by which disease passes into health, age into youth, death into life.
—Wendell Berry, *The Unsettling of America*

Eden is wounded geography.
—Sue Monk Kidd, *The Dance of the Dissident Daughter*

We are witnessing an unprecedented and massive collision between our civilization and the Earth . . . And those with the most technology have the greatest moral obligation to use it wisely.
—Al Gore, *An Inconvenient Truth*

The recovery of soils can only take place through a philosophy which sees soil fertility, not cash, as agricultural capital . . . which puts nature and human needs, not markets, at the centre of sustainable agriculture and land use. If soils and people are to live, we must stop converting soil fertility into cash and productive lands into deserts.
—Vandana Shiva, *Staying Alive*

The peasant who cuts down a tree in the Amazonian forest, the driver who empties a dump truck filled with toxic waste, and the fisherman who shoots a whale with a harpoon are only soldiers carrying out the orders of their generals.
—Eric Lambin, *The Middle Path*

A nation that destroys its soils destroys itself.
—Franklin Delano Roosevelt

To forget how to dig the earth and tend the soil is to forget ourselves.
—Mahatma Gandhi

I believe the great Creator has put ores and oil on this earth to give us a breathing spell. . . . As we exhaust them, we must be prepared to fall back on our farms, which are God's true storehouse. We can learn to synthesize materials for every human need from the things that grow.
—George Washington Carver

Dakota children understand that we are of the soil and the soil of us, that we love the birds and beasts that grew with us on this soil. A bond exists between all things because they all drink the same water and breathe the same air.
—Luther Standing Bear, *My People the Sioux*

I pledge allegiance to the soil
of Turtle Island,
and to the beings who thereon dwell
one ecosystem
in diversity
under the sun
With joyful interpenetration for all.
—Gary Snyder, "For All"

We have exactly enough time—starting now.
—Dana Meadows

Fill us with gladness for the work that must be done.
—Joanna Macy, *World as Lover, World as Self*

Human beings must take care of the garden, must take care of creation, must keep the earth. . . . In ancient Hebrew culture this is expressed in a number of ways. One of these is the bal tashchit *or "do not destroy." It is out of respect for the earth and for the Creator that this principle is given. We must work to keep the earth. This is the earthkeeping principles. . . . We are here to take joy in the beauty of this world and to keep it beautiful. . . . It must have geraniums and potatoes. It must have heifers and humming birds!*
—Calvin DeWitt,. *Inspirations for Sustaining Life on Earth*

We are the first generation with the tools to understand changes in the earth's system caused by human activity, and the last with the opportunity to influence the course of many of the changes now rapidly underway.
—Peter Vitousek

As we turn every corner of the Narrow Road to the Deep North, we sometimes stand up unawares to applaud and we sometimes fall flat to resist the agonizing pains we feel in the depths of our hearts. There are also times when we feel like taking to the road ourselves, seizing the raincoat lying nearby, or times when we feel like sitting down till our legs take root. . . .
—Matsuo Bashō

2

Air

Take a big breath. From the time our umbilical cord is severed, we breathe between 18,000 and 20,000 times a day, inhaling about 5,000 gallons of air. Air enters our lungs and quickly passes to our bloodstream. If cells don't get oxygen, they die.

Thanks to the first tiny organisms on our planet, prokaryotes, we've enjoyed a level of 21 percent oxygen in our atmosphere for over a billion years, writes Brian

Swimme in *The Universe Is a Green Dragon.* He continues, "If the concentration of oxygen were increased by only several percentage points, the conditions would become such that a single lightning strike could turn an entire forest, an entire continent, into flames. On the other hand, if the concentration of oxygen were significantly lower than its present level, we would not have the large supply of chemical potential energy necessary for advanced forms of animal life." Who knew we all depended on such tiny creatures to keep us breathing!

There's a lot about Mother Earth we haven't paid attention to. For instance, how in the world did climate change sneak up on us like this? To be sure, there are some of us still around who actually read Rachel Carson's *Silent Spring* back in 1962 when it was published. But after a ban by the EPA at the end of 1972, DDT, the World War II insecticide developed to kill insect-borne human diseases, mostly went away. We smiled when we heard about eagle's eggshells growing stronger again, but then we slipped back into decades of apathy. Now we're producing pesticides at a rate more than 13,000 times faster than we did when Carson first wrote about them. Furthermore, the Environmental Protection Agency considers 30 percent of all insecticides, 60 percent of all herbicides, and 90 percent of all fungicides to be carcinogenic. Why are we still poisoning ourselves by spending about $7 billion on 21,000 different pesticide products each year? In part, it may be because we think our lawns should look like golf courses. And we've forgotten how else to farm.

In retrospect, we should have paid closer attention to the warnings in the canvases of nineteenth-century naturalists and artists Thoreau, Muir, Marsh, and the Hudson River School of painters. Before them, the voices of Hildegard of Bingen, Meister Eckhart, John of the Cross, and the Taoists begged us to value our interwoven earth, to take responsibility, to become caring humans. About thirty years ago, scientists as well as fishermen noticed something *serious* was happening. Anecdotal stories were shared and reports issued; debates ensued. Studies were commissioned. And denounced. Today, the World Health Organization estimates that three million people die each year from air pollution. That's serious.

But with the advent of green-collar jobs, it seems that the pendulum may now be swinging in the other direction. More people than ever are "thinking green." However, we may be in danger of trivializing "green" because we're so oversaturated. All too often, serious "green" morphs into "green-lite" marketing campaigns that invite consumers to buy higher-priced items they may or may not actually need, just because they're "green."

But lest we yawn ourselves back into *jaded* (so to speak) apathy, we need to remember that there are many places on our planet where the air remains foul. The Blacksmith Institute reports that two of the ten most polluted cities in the world are in China; in Linfen, people can't hang their laundry out because in just a few hours it will be covered with soot.

When I grew up in rural Wisconsin, *pollution* wasn't in my vocabulary. I had plenty of healthy home-cooked food. Air was clean. I knew that sometimes air carried diseases; still I felt safe. After all, I was one of the first kids to receive the polio vaccine. When I was two years old, the entire world's supply of chemical antibiotics was thirty-two liters of penicillin isolated from a mold. It was first used commercially around my fifth birthday. At the turn of this new century, we were manufacturing 50 million pounds of penicillin a year in the United States alone. Then, around 2000, the first staph strain that was resistant to all known forms of antibiotics appeared. In fact, a new generation of bacteria occurs every twenty minutes. When these clever bacteria-beings encounter an antibiotic, they quickly learn how to bypass it. Meanwhile, higher temperatures will unleash not just more methane, but more bacteria—and who knows what diseases have been asleep under the ice. Gaia's temperature is heading toward levels not experienced for millions of years, and the Arctic Ocean may likely be ice-free during the summer as early as 2020, if not before.

My granddaughter, Josie, tiptoes across this earth tightrope. When she's her mother's age, there will likely be no fossil fuels left. She'll no doubt wonder why we adults didn't do something sooner.

The phrase "global warming" doesn't begin to cover the climatic disruptions we've felt over just the past several years. Some call it "global weirding." It causes dramatic droughts, hurricanes, monsoons, tornadoes, and extreme thunder and lightning displays. Fires. Floods. Water in downtown Cedar Rapids, Iowa, for instance, recently rose more than thirty feet above sea level. It not only broke all records, it exceeded the city's last flood stages by six feet!

American cars burn up about one third of all the gasoline that exists on the planet. And when we look in our rearview mirrors, we see the Indian and Chinese car markets quickly accelerating right behind us. For every gallon of gas we burn, we spew out nineteen pounds of carbon dioxide that melts our ice caps and that will, in turn, raise our ocean's levels and change their temperatures. But cars have other by-products: bad stuff comes from our brakes and tires; road chemicals run off into our groundwater. While we try to offset our emissions and reduce our carbon footprints, we are beginning to realize that we desperately need alternative solutions.

How can we better manage the carbon? Jeffrey Sachs, in his book *Common Wealth,* gives us six ways:

1. Slow or stop deforestation.
2. Reduce emissions from electricity production.
3. Reduce emissions from cars.
4. Clean up industries, such as steel, cement, refiners, and petrochemical plants.
5. Economize on electricity by being more efficient.
6. Convert point-source emissions in buildings (like furnaces) into electricity-based systems.

Imagine a green earth where pollution, carbon footprints, and lack of food and water no longer exist. A planet where people once more live *with* nature instead of *against* it. Take a deep breath and imagine.

Green Voices in Concert

When you inhale, you breathe in all the sufferings of beings and dissolve them into yourself, and when you exhale, you breathe out your own happiness, and cause happiness to others. This is just a visualization, but it builds great inner strength.
—The Dalai Lama

I picture the earth with its vapor mantle as a huge living organism involved in an increasing in- and out-breathing.
—Johann Wolfgang von Goethe

With every breath we take, God is again pumping into our lungs his exhalation of the breath of life, just as he did for Adam. If he withdrew his breath from the bubble of our world, it would instantly collapse.
—Virginia Stem Owens, "On Praising God with Our Senses"

I add my breath to your breath, that our days may be long on the Earth.
—Pueblo prayer

The breath of the air makes the earth fruitful. Thus the air is the soul of the earth, moistening it, greening it.
—Hildegard of Bingen

In Hawaii, healers are Kahna Ha, "Masters of the Breath." Some Kahunas learn to store healing energy in the heart, then project it through the laying of hands. "Aloha" means love—meeting face to face— alo, *of the breath of life—*ha.
—Kenneth S. Cohen, The Qigong Center

Let's lie down here . . . on the open prairie, where we can't see a highway or a fence. Let's have no blankets to sit on, but feel the ground with our bodies, the earth, the yielding shrubs. Let's have the grass for a mattress, experiencing its sharpness and its softness. Let us become like stones, plants and trees. . . . Listen to the air. You can hear it, feel it, smell it, taste it . . . the holy air—which renews all by its breath.
—John Fire Lame Deer

Rather than make detailed battle plans for the future, wisdom puts its effort into expanding general, adoptive options. For instance, an Earth with an intact ozone layer has more options than one without.
—Stewart Brand, *The Clock of the Long Now*

Lung-gom, *a Tantric discipline which permits the adept to glide along with uncanny swiftness and certainty, even at night, is translated "Wind-Concentration." Wind or Air in Sanskrit is* prana—*the vital energy or breath that permeates all matter.*
—Peter Matthiessen, *The Snow Leopard*

If you had a globe covered with a coat of varnish, the thickness of that varnish would be about the same as the thickness of the Earth's atmosphere compared to the Earth itself.
—Carl Sagan

Air is the mix of gases that make up the earth's atmosphere. The atmosphere is the gas layer that surrounds the planet, mixes with the planet's water, and penetrates into the soil. Air pollution comes from a change in the mix of gases, particulates, or water-based vapors that interferes directly or indirectly with human health, comfort, or safety and can eat away at metal and stone.
—Paul Hawken, *Blessed Unrest*

We need to do CPR on our planet: Conservation, Protection and Restoration.
—David Brower, Sierra Club

Chernobyl and the contamination of the Arctic graphically illustrate the global nature of air. Air is not a national or a local resource but a global commons into which we contribute our wastes and from which we draw air to fuel our bodies.
—David Suzuki, *The Sacred Balance*

There's a cost for living on this earth. We either pay now or our grandkids pay later, and I believe we should pay as we go. We've screwed up our atmosphere because everyone's had a free ride. I believe in user's fees. If you want to trash the atmosphere, you got to pay the price. I also think you should be rewarded for doing things right.
—Bernie Karl, quoted in *Earth: The Sequel*, Miriam Horn and Fred Krupp

Breast milk is supposed to be a gift. It isn't supposed to be poison.
—Eric DeWailly, quoted in *Silent Snow*, Marla Cone

At some point—I don't remember when exactly—the idea of biomagnification was introduced. This was Rachel Carson's big point, of course—that long-lived toxic chemicals, such as chlorinated pesticides, do not remain diluted when they are broadcast out into the environment. Instead, they magnify—are concentrated—inexorably as they move up the food chain. Smelt to mackerel. Mackerel to tuna. Tuna to man.
—Sandra Steingraber, *Having Faith*

You can imagine global warming this way: all those pools of oil and beds of coal beneath our feet are being drilled and dug. Emptied. For a brief moment, the resulting energy burns and does something useful: moves your car, heats your shower. But after that instant of combustion, most of the carbon in the coal or oil mixes with oxygen in the air to form the gas carbon dioxide, which drifts into the atmosphere. . . . The molecular structure of carbon dioxide traps heat from the sun that would otherwise radiate back out to space. That's all global warming is—the gaseous remains of oil fields and coal beds acting like an insulating blanket.
—Bill McKibben, *Deep Economy*

Breathing is an act of prayer.
—Frank Waters

Carbon dioxide, exhaled from our cars and our power plants like a long, collective sigh, traps heat. . . . Once trapped inside the Earth's thin layer of atmosphere, these jittery little bundles of heat energy from the Sun have to go somewhere. Many of them are going into the ocean, creating more frequent and more intense occurrences of El Niño, the huge pool of warm water that forms in the Eastern Pacific and plays havoc with the planet's weather.
—Alan AtKisson, *Believing Cassandra*

Each of us is responsible to all others for everything.
—Fyodor Dostoyevsky

Fortunately, we are finally coming to understand that the wealth of the nation resides in its air, water, soil, forests, minerals, rivers, lakes, oceans, scenic beauty, wildlife habitats, and diversity.
—Gaylord A. Nelson, founder of Earth Day, quoted in *Solutions for an Environment in Peril*, ed. Anthony B. Wolbarst

Climate change places us on a much larger stage in time, as our actions are now part of the earth's longer-term climate process itself.
—Joanna Rogers Macy

It will all come down to ethics in the end . . . how we value the natural world in which we have evolved and how we regard our states as individuals.
—E. O. Wilson

Every part of this earth is sacred to my people. Every shining pine needle, every sandy shore, every mist in the dark woods, every clearing and humming insect is holy in the memory and experience of my people. . . . All things share the same breath.
—Chief Sealth (Seattle)

Ethics is not just a matter of following abstract rules, doing what an authority figure tells you to do. Living ethically is a matter of building a healthy, whole, integrated life. . . . It is a matter of being able to look back after decades of engagement with others and say truthfully, "That was well done."
—Deane Curtin, *Environmental Ethics for a Postcolonial World*

Diversity, sustainability and equity: these are the building blocks of an environmental ethic in the making.
—Ramachandra Guha, *How Much Should a Person Consume?*

The population explosion is our largest and most worrisome and demanding human problem. If we handle it badly, then there is no sense in worrying about the others. We must use this problem as the spur that forces us to face up to the question, "Under what conditions should a human life be conceived?" . . . As the new crucial, complex decisions confront us, we must ask not only What is the right thing to do? *and* What set of values do we use? *But also* Who should be making the decisions?
—Leroy Augenstein, *Come, Let Us Play God*

Here's the new rule: break the wineglass, and fall towards the Glassblower's breath.
—Rumi

Ethical "ground" can shift dramatically over time both within and across cultures. . . . Ethics are relative. Integrity is finding and acting from your personal ethical base of values and principles. Your internal ethic is your guide for action in the world.
—Barbara Shipka, *Leadership in a Challenging World*

Something inside me has reached to the place where the world is breathing.
—Kabir

We are learning that ethics is everything. Human social existence, unlike animal sociality, is based on the genetic propensity to form long-term contracts that evolve by culture into moral precepts and law.
—E. O. Wilson, *Consilience: The Unity of Knowledge*

3

Fire

A legendary Peruvian god-man of peace once walked through a doorway from the celestial world into ours. In his hand, he held a golden disk called "Mother Sun." The disk represented not our actual solar sun, but the great central sun, the magnificent creative force of our galaxy. For years, the disk hung in the Temple of the Sun at Cuzco. When Pizarro landed with other sticky-fingered Spaniards, the disk keepers hid it in Lake Titicaca, where it resides today. Some Andean people still believe that this powerful golden disk can cause earthquakes through its vibrations and even change the rotation of earth. It's best left hidden until we earthlings know how to use it for good.

The fiery sun, millions of times bigger than earth, powers our world. We call it Sulis, Sol, Sul, Helios. Even old King Solomon . . . *sol-amun* . . . means sun god. And Brahma is called "The Most Excellent Ray." Egyptians gave all their rulers the title "Ra" or golden sun. The first Americans taught us to think of the sun as the active father or grandfather who cushions the earth, more of a mother or grandmother. Other cultures see the sun, with its spirals and wheels, as a feminine symbol, bringing rebirth and renewal. People still wear amber and gold to remind themselves of the sun's radiant power.

When the Ancients spoke of the four elements, they usually listed fire first. Fire represents intuition and inspiration—those god-gifts Prometheus gave to humans at the expense of having his liver torn out each night. (Zeus, apparently, didn't take kindly to Prometheus's sneaking fire away from the sun to benefit humans.) Teilhard de Chardin, a contemporary Prometheus, predicted that one day we would all burst into a new consciousness based on love, and we would experience fire for the second time. Our hearts glow red-hot just thinking of that possibility.

Meanwhile, we use up the sun's gifts at an alarming rate. All our fossil fuels lie hidden under earth's crust because the sun once warmed the plants that produced them. Thom Hartmann, who calls fossil fuels our ancient sunlight, reminds us that everything on our planet is derived from absorbing the energy of light. Everything. But hidden, like that golden disk deep in the lake, we still have Prometheus's gift of inspiration—and many people are using it to create alternative energies. Take the Toyota Prius. The Japanese decided they wanted to create something totally "out of the box." A whole new system. So they asked themselves, "What if a car could make some of its own energy as it runs? What if a car going downhill could store energy in a battery to power going uphill? What if every time the driver put on the brakes, she created more energy?" If American auto companies had asked themselves similar questions, they could have been more competitive and less prone to bankruptcy. Cars will continue to be manufactured, and by 2020 our planet will have doubled its present number of automobiles to one billion vehicles! But getting around on dwindling fuel is only one of the many befuddling challenges we face.

In spite of our adult grab-at-everything tendencies, kids all across the world are becoming greener and greener. Be prepared to discuss your carbon footprint next time the grandkids are over. They're savvy. In a recent survey conducted by the Nickelodeon television network and the Pew Center on Global Climate Change, kids ages eight to fourteen ranked environmental problems as among the five most important issues facing the world today. One third of the young respondents said they believe individuals have the greatest responsibility to protect the environment.

If we have hands and a will, we have the power to act. We need not wait for someone to encourage us—or even to show us the way. We know how. The Dalai Lama has warned us that negative actions always bring about suffering, while positive actions always bring happiness. The key word: *always.*

The Hindu teacher Vivekananda posed the following questions: How much more do I need to be happy? How much less can I have and still be happy? We live in fear of running out of things. Believing there is a pie big enough to go around is at the heart of abundant living. The Swedes have a word that probably comes out of their long history of drinking ale by passing around a large horn: *Lagom. Lag* means "team" and *om* means "around." In other words, confidently pass the beer around to your teammates, and rest assured that there will be enough for everybody. My son Joel, who has lived in Sweden for many years, assures me the practice of *Lagom* still appears at Swedish dessert time. "If there's a last piece of cake," he says, "any self-respecting Swede will cut it in half and then the next person cuts that in half and so on until you need a microscope and a scalpel. . . ."

Understanding our unequal parceling out of the world's food supplies means we need to understand more about complicated policies. And, obviously, wars do not help the food, water, or health crises.

Thoreau asked two key questions—questions we'd be wise to ask ourselves every day. How much is enough? How do I know what I really want? If we answer those two questions honestly, then we'll remember what we are: fire-beings waiting to burst into conscious flame for the second time.

Green Voices in Concert

O Agni, Holy Fire! Purifying Fire! You who sleep in the wood and ascend in shining flames on the altar, you are the heart of sacrifice, the fearless wings of prayer, the divine spark hidden within everything, and the glorious soul of the sun.
—Vedic hymn

There are those who would set fire to the world.
We are in danger.
There is time only to move slowly.
There is no time not to love.
—Deena Metzger

God appears to you not in person but in action.
—Mahatma Gandhi

If the world were a town of 1000 people there would be 564 Asians, 210 Europeans, 86 Africans, 80 South Americans and 60 North Americans. 700 people would be illiterate and 500 would be hungry. About 250 people would be consuming 70% of the energy, 75% of all metals, and 86% of the wood.
—Margaret Lulic, *Working As If Life Mattered*

When spider webs unite, they can tie up a lion.
—Ethiopian proverb

Chindogu *is a Japanese word for all the useless things we might be tempted to buy—windshield wipers for your spectacles . . . slippers with mops underneath so that you can polish the floor as you walk around the house. . . . Technology thrives on* chindogu.
—Charles Handy, *The Hungry Spirit*

Don't confuse activity with effective action. We don't have time to get lost or distracted.
—Betsy Taylor

We live in a time when the greatest form of courage is to act as if our lives made a difference.
—William Sullivan, *The Secret of the Incas*

If we cannot see the wholeness in the world then we cannot take actions which are consistent with the wholeness which exists.
—David Bohm, *Fragmentation and Wholeness*

First they ignore you, then they laugh at you, then they fight you, then you win.
—Mahatma Gandhi

Buddhists sometimes speak of hungry ghosts. By this, I imagine they mean beings that are constantly consuming but are so insubstantial, so ghostly and removed from living, that they gain nothing from that consumption, so they go on consuming. We live in a time ruled by hungry ghosts.
—Richard Manning, *Inside Passage*

When we had a lot of community and not much stuff, it made sense that we aimed for stuff. But why do we keep aiming for it? Why don't we realize we have enough, and turn our attention elsewhere?
—Bill McKibben, *Deep Economy*

They make the garbage in the south
And ship to the north
The county budget pays the bill
To truck it back and forth.
—Kent Gregson

We tend to sit back and point accusing fingers at the major industrial polluters of the world, at the developers, at all who are harming the environment, and at governments that fail to take a tough stand against them. We ourselves become overwhelmed by the magnitude of the task of restoring harmony once again between ourselves and the natural world. . . . [we think] "I am just one person, what I do and do not do cannot possibly make any difference." . . . Suppose we turn this "just-me-ism" around. Suppose all the millions of us who care realized that what we do each day does make a difference—because our actions will be magnified millions of times over. . . . We are consumers, and in a consumer-driven society we have enormous power to effect change—if we act together.

—Jane Goodall, *Solutions for an Environment in Peril*

A responsible consumer is a responsible conservationist.

—Wendell Berry, *The Unsettling of America*

Reduce, reuse, recycle, retrieve, redeem.
—Richard Powers, *The Echo Maker*

Trash is the visible interface between everyday life and the deep, often abstract horrors of ecological crisis.
—Heather Rogers, *Gone Tomorrow: The Hidden Life of Garbage*

It really isn't garbage till you mix it all together.
It really isn't garbage till you throw it away.
Just separate the paper, plastic, compost, glass and metal
And then you get to use it all another day.
—Elizabeth Royte, *Garbage Land*

In the real world, the very necessary task of recycling is at best calisthenics for the marathon we must run. Realism, sadly, demands that we recognize the need for deep and fundamental change—for recycling our cars into buses and bicycles.
—Bill McKibben, *Hope, Human and Wild*

The stuff we set out on the curb is circling back to bite us. We burn our electronic waste and its chemical fallout shows up in the breast milk of Eskimos and in the flesh of animals we eat. We bury our household waste and poisons rise into our air and leech into our waterways. . . . Plastic is Satan's resin. It doesn't go away.
—Elizabeth Royte, *Garbage Land*

We need to make a world-scale "natural contract" with the oceans, the air, the birds in the sky. The challenge is to bring the whole victimized world of "common pool resources" into the Mind of the Commons. As it stands now, any resource on earth that is not nailed down will be seen as fair game.
—Gary Snyder, *Practice of the Wild*

At first people refuse to believe that a strange new thing can be done, then they begin to hope it can be done, then they see it can be done—then it is done and all the world wonders why it was not done centuries ago.
—Frances Hodgson Burnett, *The Secret Garden*

Corporations today are controlling our lives the same way the British controlled life in India, and I'm basically using Gandhi's methods to fight them. His vision was that a self-reliant population could throw off British rule nonviolently. So he advised people to grow their own food and make their own clothes.
—Judy Wicks, "Table for Six Billion, Please," quoted in *The Sun*

This is a time to be still no longer . . . a time for crying out, as Hebrews cried out in bondage and Jesus on the cross. . . . We need to give vent to our massive pain and fear. A people must move from muteness to outcry if it is ever going to take the next step.
—Harvey Cox

Do the easy things, then do a few more challenging things that we really believe in and enjoy.
—Alex Steffen, *Worldchanging: A User's Guide for the 21st Century*

Today, I see a new kind of activist emerging. Not one who is angry or burned out, but one whose belief that things can be different goes deeper than a passing optimism. We've had plenty of very sophisticated analysis of what's wrong with the world, much of it quite helpful. But what's often been missing is the vision to help people connect the desire to change their lives with a commitment to change their communities. That vision will likely be rooted in moral and spiritual values.
—Jim Wallis, *Faith Works*

We need a movement. We need a political swell larger than the civil rights movement—as passionate and as willing to sacrifice. Without it, we're not going to best the fossil fuel companies and the automakers and the rest of the vested interests that are keeping us from change.

—Bill McKibben, *Yes!* Magazine

Perhaps all those who work for change should study jujitsu, learning in their very bodies to work with the impulses and directions that are already there, rather than opposing them.

—Mary Catherine Bateson, *Peripheral Visions*

Sometimes we achieve the impossible sooner than we expect. Knowing that can stiffen our resolve. But it can also tempt us to place too much emphasis on outcomes; it can cause us to become unduly impatient, brittle, setbacks easily breaking our will. A deeper, more farseeing hope, by contrast, combines realism with resilience, acknowledging terror and suffering without giving in to them.
—Paul Rogat Loeb, *The Impossible Will Take a Little While*

Culture is not about what is absolute, real or true. It's about what a group of people get together and agree to believe. Culture can be healthy or toxic, nurturing or murderous.
—Thom Hartmann, *The Last Hours of Ancient Sunlight*

There is vitality, a life force, an energy, a quickening that is translated through you into action, and because there is only one of you in all time, this expression is unique. . . . There is only a queer, divine dissatisfaction, a blessed unrest that keeps us marching and makes us more alive than the other.
—Martha Graham and Agnes de Mille, *Dance to the Piper*

It's not what vision is, it's what vision does.
—Peter Senge, *The Necessary Revolution*

In this battle I have understood that working for a clean environment today is working towards peace for humanity tomorrow . . . facing the future. That is what I intend to do.
—Pablo Fajardo

There are all different kinds of voices calling you to all different kinds of work, and the problem is to find out which is the voice of God rather than of Society, say, or the Superego, or Self-Interest. . . . Neither the hair shirt nor the soft berth will do. The place God calls you to is the place where your deep gladness and the world's deep hunger meet.
—Frederick Buechner, *Wishful Thinking: A Seeker's ABC*

We must learn to step forward without the grand plan and without overcoming all possible objections.
—Margaret Lulic, *Working As If Life Mattered*

Every week counts. How terrible and shamefully bad conditions will be in the twenty-first century, or how far down we fall before we start on the way back up, depend upon what YOU and others do today and tomorrow. There is not a single day to be lost. We need activism on a high level immediately.
—Arne Naess, "Deep Ecology for the Twenty-Second Century," quoted in *The Trumpeter*

We cannot wait any longer to solve this crisis. We have nearly all the tools we need to solve this problem, perhaps with one exception. What we are missing is the political will that would be required to really affect change. Thankfully, in a democracy like ours, political will is a renewable resource.
—Al Gore, *Earth in the Balance*

Never underestimate the power of groups of committed citizens to change the world. In fact, it is the only thing that ever has.
—Margaret Mead, "Our Open-Ended Future"

You must believe as if your every act, even the smallest, impacted a thousand people for a hundred generations. Because it does.
—Thom Hartmann, *The Prophet's Way*

After the sun's energy is captured by the green plants, it flows through chains of organisms dendritically, like blood spreading from the arteries into networks of microscopic capillaries. It is in such capillaries, in the life cycles of thousands of individual species, that life's important work is done. . . . The study of every kind of organism matters, everywhere in the world.
—Howard T. Odum, *Environment, Power, and Society*

Though there are no dragons, we are dragon fire. We are the creative, scintillating, scaring, healing flame of the awesome and enchanting universe. . . . Are we tending and revering this fire?
—Brian Swimme, *The Universe Is a Green Dragon*

The world has enough for everybody's need, but not enough for everybody's greed.
—Mahatma Gandhi

The conclusion is always the same: love is the most powerful and still the most unknown energy of the world.
—Pierre Teilhard de Chardin

All this world is strung on me like jewels on a string. I am the taste in the waters, the radiance in the sun and moon, the sacred syllable Om *that reverberates in space, the manliness in men. I am the pleasant fragrance in earth, the glowing brightness in fire, the life in all beings.*
—Bhagavad Gita VII: 7–9

Even fish, though they are water creatures, may be restored by fire. Thus a Tlingit fisherman takes care to burn the bones, so that the same individual can be caught year after year. It never actually dies, because, as the explanation goes, "When you get that fish, it's not the real fish. It's just the picture of it."
—John Bierhorst, *The Way of the Earth*

I'd put my money on the sun and solar energy. What a source of power! I hope we don't have to wait until oil and coal run out before we tackle that.
—Thomas Edison

Without the decay of the Universe there could have been no Sun, and without the superabundant consumption of its energy store the Sun could never have provided the light that let us be.
—James Lovelock, *The Ages of Gaia*

At one end, "cleanup" means making certain that future human generations need not worry about residual contamination left over from weapons production and other nuclear activities. All the residuals have been converted into non-harmful materials, or secured so completely that future generations are extremely unlikely to come into contact with them. At the other end of the continuum, however, "cleanup" would mean securing contaminants with some combination of physical barriers and institutional controls, while waiting for either new and more efficient cleanup technologies, or new ways to re-use materials. Future generations would decide how best to further manage these materials and deal with the risk.

—Max S. Power, *America's Nuclear Wastelands*

Global warming is cheaper to fix than to ignore. Because saving energy is profitable, efficient use is gaining traction in the marketplace. . . . The climate problem was created by millions of bad decisions over decades, but climate stability can be restored by millions of sensible choices.

—Amory B. Lovins, *Scientific American*

If everyone on Earth disappeared, 441 nuclear plants, several with multiple reactors, would briefly run on autopilot until, one by one, they overheated.

—Alan Weisman, *The World Without Us*

Chernobyl turned me into a different person. . . . First, Chernobyl became a decisive test for the new policy of glasnost . . . [we] decided on the very first day to publish all the details about the catastrophe as soon as they reached us. . . . Secondly, my belief in the absolute reliability of technology was shattered. For thirty years we had been assured that "the peaceful atom was no more dangerous than a samovar." . . . Thirdly, my time-scales changed radically. The half-life of Caesium 137, the radioactive isotope most damaging to health that escaped from the "cauldron" of Chernobyl, is thirty years, which means that this element will still be poisoning foodstuffs and affecting the health of populations in the polluted areas for a long time to come. . . . And fourthly, Chernobyl stiffened my resolve to establish new international contacts and demonstrate emphatically that we are a single humanity sharing a single planet. After all, the radioactive cloud had spread around the globe within a few days, and traces of radioactivity had been found thousands of kilometers distant from the location of the catastrophe.

—Mikhail Gorbachev, *Manifesto for the Earth*

The world is at a turning point . . . that can only happen once in the evolution of our planet. . . . Right now almost the only policy-makers who talk seriously about planning for the long-term—even a few thousand years from now—are people discussing how to dispose of nuclear waste properly. They wonder, for example, how a nuclear waste dump which will be dangerously radioactive for perhaps ten thousand years, should be identified with signs to keep people from unknowingly digging it [up] when no one may read our language anymore or understand the concept of radioactive waste. . . . One proposed we should carve into stone at the sites of nuclear waste dumps copies of Edvard Munch's terrifying painting The Scream*. . . . Think cosmically, act globally.*

—Joel R. Primack and Nancy Ellen Abrams, *The View from the Center of the Universe*

We can pray to the Sun for understanding of what is happening. The Sun will help us if we ask. The Sun is not God, but a part of God. God's eye, a reflection of the eyes of God. Hunab K'u. We are a reflection of Hunab K'u, too. We are Hunab K'u—as is all of creation. Everything is integrated. Nothing is separate. The Maya have known this always. When the Maya make rituals to the Sun, they regard it as father, and ask for help. The Earth is mother. . . . We have a piece of the Sun in us, which gives us life.

—Hunbatz Men, *Profiles in Wisdom*

You must do the thing you think you cannot do.

—Eleanor Roosevelt

The flash of light in the night in the desert, which I had always thought was just a dream, developed into a family nightmare. It took fourteen years, from 1957 to 1971, for cancer to manifest in my mother—the same time, Howard L. Andrews, an authority in radioactive fallout at the National Institutes of Health, says radiation cancer requires to become evident. . . . When the Atomic Energy Commission described the country north of the Nevada test site as "virtually uninhabited desert terrain," my family and the birds at Great Salt Lake were some of the "virtual uninhabitants."

—Terry Tempest Williams, "The Clan of One-Breasted Women" quoted in *The Impossible Will Take a Little While*, ed. Paul Rogat Loeb

4

Water

In early spring, it marks the edge of our property with a ferocity that surprises me every year. Before the topography maps were crafted, someone gave it a name, no doubt for the trees that grow farther down its path: Apple Creek. We've become fast friends. I sit on a rocky outcropping above the brook and watch water leap and froth like buckets of crazed Fresca.

In winter, the ice piles up in blue layers, and I can see, just below its glacial framework, energetic movement. It presses on downed logs, around slate outcroppings, and eventually reaches our branch of the White River. Which reaches the Connecticut. Which extends to the Atlantic Ocean.

Sitting on my wooden bench a few feet above Apple Creek, I act as a guardian, recalling the Indo-European root word for "to care" is *gar.* I "gar" the water. The brook water becomes "all water" from my overlook. While comfortable with the litanies that pepper our Episcopal Common Book of Prayer, I use simpler words: "Thank you." "Be well." I don't plant prayer sticks, or hang prayer flags. In the quiet, I remember that Frederick Buechner said prayer is like shooting shafts into the dark. My words go out, like shafts or arrows. Or up. Or down into the mossy rocks. Or deeper inside me. I'm not sure where prayer-shafts land; nor do I care. Prayer's not about aiming right or hitting the mark. It's not about being helpful or impressing anyone. Nor, I remind myself, is prayer about being in control. To pray is simply *to care*—to care about my little brook and to care enough about my own little self to search out the words that express what my heart knows. Prayer acts as a thought-anchor, first revolving around me and my needs, then moving on to others. In her book *Acedia & Me,* Kathleen Norris writes about this strange medieval relic of a word, *acedia.* It has a deeper meaning than apathy, sloth, boredom, or torpor. It's a total "lack of care." This is not the time to be apathetic about each other. Or our planet. Norris quotes the Carmelite Constance Fitzgerald, who reminds us that "We see cold reason, devoid of imagination, heading with deadly logic toward violence, hardness in the face of misery, a sense of inevitability, war and death." We must name and confront our acedia before it becomes a soul-sickness from which we may not recover. We are the "gar-dians" and "care-dians" of all waters and of all land. And of each other. According to Genesis, and later good-book messages, it's the real reason we're here.

Like the somber couple standing with pitchforks in Grant Wood's *American Gothic,* farmers have stood for hundreds of years, pitchforks upraised, guarding the land. Of the nine families who now live on our little Vermont mountain, we who share in Apple Creek's largesse, only two families still farm the land. One other owns a few beef cattle, pigs, and chickens. Of our two full-time farming neighbors, it appears only one has kids who want to continue what their hardworking parents have established. The average age of American farmers nears sixty. Who will grow our food tomorrow? And will there be enough water to irrigate it? As I write this, dire predictions from California anticipate that a million acres will go without planting in

2009 because there is not enough water. Increasing fuel prices spiral many small farms into bankruptcy. In the United States we still insist on eating, on average, two hundred pounds of meat a year. Do we wish to continue to fill our refrigerators and freezers solely with meat that corporate farms pump up with hormones, antibiotics, and other sorry factory profit-driven practices? Or do we want to eat from what is nurtured by the Apple Creeks of the world?

As *gar-dians,* it's up to us to pay attention to what we eat and drink. We throw away 2.5 million plastic water bottles in the United States alone *every hour!* We've become very fussy consumers. We like our green beans to be the same size. We feel we deserve strawberries year-round and ripe mangos in December. Huge mowing machines leave cuts on spinach and salad greens where *E. coli* can grow. So growers dunk the greens into chlorine to kill the bacteria and then refrigerate them (using more fuel), only to learn that only the older forms of bacteria can be killed by cold temperatures—the newer, smarter ones can't. So bacteria continues to grow, only more slowly. Do we really need to ask why our food sometimes makes us sick?

My little Apple Creek tears up at the thought.

It's up to us to start questioning. Why, for instance, do we continue to buy things that are bubble-wrapped in so much plastic? Why do we still get so many catalogs in the mail? According to Bill McKibben, in 2007, 19 million catalogs were sent out, expending as much CO_2 as two million cars would discharge. And we wasted 53 billion gallons of water making those catalogs.

And how will the planet sustain the 80–90 million babies that are born each year? Every year we produce enough new humans to populate Canada, Australia, Denmark, Austria, and Greece. Twelve thousand babies are born each day in Pakistan alone. And the sad thing is that only one third of the people living in Pakistan have access to clear water; 15 percent of Pakistanis have sewage; only twenty-five out of every hundred will go to school. Robert Kaplan, who wrote *Ends of the Earth* about his journeys around the world, calls Pakistan a land of dry faucets and leaky toilets. And in India, a land of tremendous growth, 270 million people are continuously hungry.

In the 1970s, all of our refrigerators contained CFCs, or chlorofluorocarbons. Then, some people began to worry about a hole we'd discovered way up there in the ozone layer—a tear in our protective membrane. Two scientists, F. Sherwood Rowland and Mario Molina, figured out that chlorine was being released into the ozone layer. Chlorine, they hypothesized, was mixing with ozone in the stratosphere, and when the sun hit it, earth's fragile skin was peeling away. In *An Inconvenient Truth,* Al Gore says, "In 1984 a dramatic hole in the ozone layer was discovered above Antarctica, just as the scientists had foretold." Then twenty-seven nations signed the Montreal Protocol on Substances that Deplete the Ozone Layer, a group that now includes over one hundred eighty countries. CFC levels have stabilized or declined, and the hole in our ozone layer is actually shrinking. It's not completely recovered, but we have proved to ourselves that by working together, great things *can* be accomplished.

My bubbling Apple Creek reminds me that in the next fifty years, almost half of us will face serious drinking water shortages unless we act soon. Our oceans could rise as much as twenty feet! Of China's forty-seven principal cities, twenty-two have contaminated drinking water. We're clinging to what Stewart Brand called Life-Raft Earth. And there may not be enough inflatable orange jackets to go around. Poverty and climate change go hand in hand. It is the poor among us who are most vulnerable to natural disasters because so many live on riverbanks and steep slopes.

When people work together, like the New Yorkers who convene each year to plant daffodils in their parks, great things can happen. Harvard Business School professor John Kotter says that within six to twelve months, visible changes in the environment will always result from people who succeed at improving things. Like daffodils planted in the fall to bloom in the spring, good things are coming! I know it as surely as I know Apple Creek runs faster in spring.

When we harmonize our thinking with our actions, we reach what the ancient philosophers called "right living." Right action, right feeling, and right thinking. But how do we know what's "right"? What's right for me might not be right for you. Mother Teresa said it's not how *much* you do, but *how much love* you put into the doing that truly matters. Given a chance, our hearts will tell us what's "right"—especially if we are in dialogue with others we respect. Look around. You'll be amazed. People everywhere are engaged in building hopeful, forward-leaning examples that serve as models to us all. The rest of us are pulled along in the pounding wake of their actions.

Active people find the best rhythm for performing daily routines, and they aren't subject to other people's whims or intrusions. Seriously thinking about and planning our actions keeps us from operating on blind impulse. We "see" what we "do" before we do it. And when we visualize our actions, we are more apt to anticipate the outcomes.

Every action we take precipitates a reaction; every action involves some kind of change. We live with uncertainty, so we can't always see the results. But as Ken Carey writes in

The Starseed Transmissions, our actions have huge effects on the world around us. This is a very sobering thought. "The things you do today, the things you do tomorrow, the things you do next week, have far greater significance than you suspect. . . . Through your actions today, vast worlds will be created and destroyed."

Green Voices in Concert

The surface of the earth had not appeared. There was only the calm sea and the great expanse of the sky. . . . There was nothing standing; only the calm water, the placid sea, alone and tranquil. Nothing existed. . . . Thus let it be done! Let the emptiness be filled! Let the water recede and make a void, let the earth appear, and become solid; let it be done. Thus they spoke. Let there be light, let there be dawn in the sky and on the earth! There shall be neither glory nor grandeur in our creation and formation until the human being is made, man is formed. So they spoke.

—From the Mayan *Popul Voh*

If there is magic on this planet, it is contained in water.

—Loren Eiseley

We made from water every living thing. . . . Consider the water which you drink. Was it you that brought it down from the rain clouds or We? If We had pleased, We could make it bitter.
—*Qur'an:* 21:30, 56:68–70

Limitless and immortal, the waters are the beginning and end of all things on earth.
—Heinrich Zimmer

The oceans, those great expanses of deep blue sea, have far more to them than the mere capacity to dazzle an observer in outer space. They are vital parts of the global steam engine that transforms the radiant energy from the sun into motions of air and water which in turn distribute this energy all over regions of the world. Collectively, the oceans form a reservoir of dissolved gases that helps to regulate the composition of the air we breathe and to provide a stable environment for marine life—about half of all living matter.
—James. E. Lovelock, *Gaia*

Please tell somebody to turn on our water. We haven't had any since Sunday.
—A hand-lettered sign in the Belize countryside

There is enough water for everyone on the planet, but the trick is how to get it and make it available for every use and every user.
—Aly M. Shady, *Resource*

Once I lived in a town called Manitou, which means "Great Spirit," and where hot mineral spring water gurgled beneath the streets and rose up into open wells. I felt safe there. With the underground movement of water and heat a constant reminder of other life, of what lives beneath us, it seemed to be the center of the world.
—Linda Hogan, *Dwellings*

Of all our natural resources, water has become the most precious. . . . In an age when man has forgotten his origins and is blind even to his most essential needs for survival, water along with other resources has become the victim of his indifference.
—Rachel Carson, *Silent Spring*

The surface of the earth is . . . 71 percent water. Just like my very body . . . [t]wo hydrogen and one oxygen, molecules in concert, a fluidity around and within that defies the solid life, undercuts the unchangeable with motion, washing away and re-creating, over and over.
—Marybeth Holleman, *The Heart of the Sound*

We are water—the oceans flow through our veins, and our cells are inflated by water, our metabolic reactions mediated in aqueous solution . . . a sacred liquid that links us to all the oceans of the world and ties us back in time to the very birthplace of all life.
—David Suzuki, *The Sacred Balance*

Ocean energy can contribute a great deal toward the protection of our atmosphere—without damaging marine ecosystems that are equally vital to the planet's future.
—Fred Krupp, *Earth: The Sequel*

"The [Platte] river's being used up. Fifteen dams, irrigation for three states. Every drop used eight times before it reaches us. The flow is a quarter of what it was before development. The river slows; the trees and vegetation fill in. The trees spook the cranes. They need the flats. . . . They're brittle—a low annual recruitment rate. Any large habitat break will be the end. . . . A few more years, and we can say goodbye to something that's been around since the Eocene."

Something in Daniel mourned more than the cranes. He needed humans to rise to their station: conscious and godlike, nature's one shot at knowing and preserving itself. Instead, the one aware animal in creation had torched the place.

—Richard Powers, *The Echo Maker*

When I see people . . . waste their money buying bottled water at the vending [machine] when it's standing right next to a water faucet, you really have to wonder at the utter stupidity and the responsibility sometimes of American consumers.

—Rocky Anderson, Grist.org

Muslims, whether they are nomadic Bedouins in Jordan or Darfur or residents of low-lying Bangladesh, are among the most immediate victims of climate change. Prolonged droughts and in other places floods are forcing these climate refugees to abandon ways of life that have sustained these societies for thousands of years. The Intergovernmental Panel on Climate Change (IPCC) has suggested 150 million environmental refugees will exist by 2050.

—Juliette Beck, *Yes!* Magazine

The question is, when do we all become indigenous people? When do we become native to this place? When do we decide we are not leaving?
—William McDounough, *FEED* Magazine

Living in harmony with the peace and quietude of nature taught the Seneca self-discipline. They moved slowly, spoke softly, and developed a natural quiescence. This silence had to be learnt and signified perfect harmony in spirit, mind and body. To master this characteristic meant functioning harmoniously within one's immediate environment.
—Twylah Nitsch, *Language of the Trees*

The ice is melting. We know this, and we also know that there will be floods and we know that it will affect us all. When the ice in Greenland melts and these enormous amounts of water flow into the ocean, then the balance of ocean currents will change. The Gulf Stream, which is responsible for Europe's temperate climate, could vanish and possibly bring about a new ice age. . . . But when will they wake up? Will it be when the first floors of their skyscrapers have been flooded? Or the fifth floor?

—Angaangaq Lyberth, *Icewisdom.com*

The first time I heard about global warming, I thought, I don't believe those Japanese. . . . Well, they had some good scientists, and it's become true.

—Morris Kiyutellus, a lifelong resident of Shishmaref, Alaska, in *Field Notes from a Catastrophe*, Elizabeth Kolbert

The prophecy of my people was that someday the great Sea Eagle would descend from the mountains of ice and that she would carry with her a fresh scent of knowledge combined with wisdom. After melting the ice in the heart, who you are—the real beauty and wisdom—will shine forth naturally. When you are trained to hear what your heart is saying and you live out its messages, you will see with eyes of faith into the future and see the difference. Unless this journey within is made and the distance conquered man will not realize his immensity and learn to soar like the eagle. Melting the ice in the heart of man means that the world we live in will have a chance to change.
—Angaangaq Lyberth, *Icewisdom.com*

Do what we will, the Colorado will one day find an unimpeded way to the sea.
—Donald Worster, *Under Western Skies*

We're trying to change the climate-system—to avoid the unmanageable and manage the unavoidable! We are trying to affect how much the rain falls, how strong the winds blow, how fast the ice melts.
—Thomas L. Friedman, *Hot, Flat, and Crowded*

Big dams did not start out as a cynical enterprise; they began as dream. They have ended up as a nightmare. It's time to wake up.
—Arundhati Roy, *Art India, Inc.*

The ice in some places on the coast is now melting four times faster than before.
—Abbas Khan

In the United States, whose 5,500 large dams make it the second most dammed country in the world, we have stopped building large dams, and are now spending great amounts of money trying to fix the problems created by the existing ones.

—International Rivers Network

Woody Guthrie was hired by the federal government in 1941 to write and sing songs chronicling the building of the dams. Dylan, the workingman's poet turned propagandist hired to sing "Roll on, Columbia, roll on," to a river whose rolling was about to be locked behind concrete walls. . . . Something has died with the salmon . . . they embody the totality of the life of the place. This life is slow, motion, exchange.

—Richard Manning, *Inside Passage*

We always had plenty; our children never cried from hunger, neither were our people in want.... Our village was healthy.... If a prophet had come to our village in those days and told us that the things were to take place which have since come to pass, none of our people would have believed him.

—Black Hawk, Sauk and Fox Chief, *Autobiography of Black Hawk*

When we burn grass for grasshoppers, we don't ruin things. We shake down acorns and pinenuts. But the White people plow up the ground, pull down the trees, kill everything.... They blast rocks and scatter them on the ground.... How can the spirit of the earth like the White man?... Everywhere the White man has touched it, it is sore.

—A Wintu woman of California, quoted in *Freedom and Culture*, Dorothy Lee

Indigenous science teaches that all that exists is an expression of relationships, alliances and balances between what (for lack of better words) we call energies, powers or spirits . . . [for the Blackfeet] the circle is always left open so that the new may enter. Nothing is permanent, no situation is ever fixed, and no category is ever closed.

—F. David Peat, *Blackfoot Physics*

Hopi land is held in trust in a spiritual way for the Great Spirit, Massau'u. . . . The Great Spirit has told the Hopi Leaders that the great wealth and resources beneath the lands at Black Mesa must not be disturbed or taken out until after purification when mankind will know how to live in harmony among themselves and with nature. . . . It is unthinkable to give up control over our sacred lands to non-Hopis.

—Thomas Banyacya, "Statement of Hopi Religious Leaders"

The Inuit mark the cycle of seasons . . . with the life-affirming arrival of the beloved caribou herds to their lands in May. . . . In this way, the Inuit pay homage not only to primary celestial rhythms in their universe, but to the more subtle and imprecise ecological rhythms in the visible patterns of life, experience, and memory upon the surface of the land.

—David Suzuki and Peter Knudtson, *Wisdom of the Elders*

Do you want to improve the world?
I don't think it can be done.
The world is sacred.
It can't be improved.
If you tamper with it, you'll ruin it.
If you treat it like an object, you'll lose it.

—Lao Tzu, *Tao Te Ching*

5

Time

When I was a toddler, my dad left Wisconsin to help build what was then called the Al-Can Highway, The Trail of '42. I was too young to know why he wasn't around, but letters arrived from strange places like Whitehorse, Fort Nelson, and Skagway. I value the pictures of him standing near heavy equipment in muddy ruts, or next to someone scraping a moose hide. With the Army Corps of Engineers, he worked on a 1,400-mile road. Had this road begun in

Boston, it would have stretched to the middle of Iowa. It crawls through Dawson Creek, across British Columbia, over the Yukon Territory, and up to Delta Junction, Alaska. They built it because our troops needed a way to move supplies and defend our borders from the Japanese who, after Pearl Harbor, threatened our country.

Now that road is a shiny ribbon. Back then, it was a muddy track running through wilderness, mosquitoes and black flies, deep snows and nearly impassable terrain over elevations of around 4,000 feet, across muskeg and melting permafrost. That stretch of highway shows what can happen when people use keen negotiating skills to reach a common plan.

On February 2, 1942, President Roosevelt formed a special cabinet committee to come up with a viable way to build a road to Alaska through Canada. By February 14, they'd decided on the most feasible route, and Roosevelt issued a directive for the work to begin. Negotiations with the Canadians gave the United States the right-of-way; Canada waived import duties, sales taxes, income taxes, and immigration regulations and permitted Americans to take timber, gravel, and rock from along the route. In return, the United States agreed to pay for the construction and turn over the Canadian portion of the road to the Canadian government six months after the war ended.

The first troop train arrived at Dawson Creek on March 2, and the bulldozers headed north. Racing against spring thaws, working day and night, they carved three or four miles through the wilderness each day and then moved their tents and set up a new camp.

The road was built in eight months and twelve days. When we wonder if we can create alternative energies or high-speed railways, grids to carry solar and wind energy, we should remember the Al-Can. "Yes we ALL can!"

We need new models for many things, because it's obvious that the old ones don't work anymore. Our whole civilization needs a jumpstart. James Lovelock once suggested that we compile a manual for restarting civilization, beginning with how to make fire and progressing right up to genetic cloning. What a huge cross-disciplinary assignment that would be! Where and how would it be kept? We could store the knowledge in seed bunkers like the one 600 miles from the North Pole, set deep inside an Arctic mountain cave in Svaalbard, Norway, that will hold 4.5 million seed samples from food crops around the world—just in case. Or we could carve the information on stone, as the Buddhists did to guard their ancient prayers. Or perhaps we could tuck it away in dry caves on wood-covered fold-out codices as the early Mayans did. How can we assume a computer somewhere in the future could be fired up and show its hidden data flowing along like my ice-encrusted Apple Creek water?

I sense it's time we all assume our "elder" roles, regardless of our actual ages. Elders are people who remember. They are the ones who watch, who keep track, who dare speak out, and who have the courage to act. Rather than wait for other people to tell us what to do, let's all be elders in the making. Right now, in this moment.

We all know the truth, as ancient people knew, that if we act responsibly, the planet will be healed. But it's ultimately up to us to keep our eyes focused on the horizon rather than on our feet, looking to what Stewart Brand calls "the clock of the long now." It's up to us elders to keep asking the Parsifal grail questions: *What really ails us? Whom do we serve? And what can we do today to benefit tomorrow?*

But it will take a fair amount of faith. As a wise friend named John once told me, faith is not nebulous. Faith is not transparent. Faith is absolute. It's the action, he said, that allows the spirit and the body and the mind to unite. Faith then becomes a light in the darkness and shows the way. What value, he asked, is awareness if people do not apply it in some form or other?

We humans have done some pretty remarkable things in the past besides building that northern highway in eight months. Medieval cities used to be cesspools. In fact in Paris streets were appropriately enough named "Merde." And remember when the Cuyahoga River was awash with flames? Or when every breath drawn in London burned the throat?

Cleanups are never simple or easy. And they're certainly not inexpensive. While trying to avoid icebergs, the *Exxon Valdez* hit a reef in 1989 and spilled almost 11 million gallons of crude oil into Prince William Sound off the coast of Alaska. When Chernobyl popped in 1986, the Ukraine, Russia, and Belarus lost over 2.5 million people. And in 1998 when a mining reservoir burst in Spain, it seeped toxic sludge and heavy metals into several precious rivers. Cleanups from these disasters cost billions—maybe trillions as we measure money now.

There's an African proverb that goes like this: *You can count the apples on a tree but you cannot count the trees in an apple.* Like the apple, we are all seeded with possibilities. But making choices always calls for sacrifice. For instance, if you want to be a professional athlete, your amateur status must be sacrificed. If you want to build cathedrals, you'll have to stop building ranch houses. We're feeling the tension-tugs of what exists and what could exist. And what we'll have to give up. Which road in the yellow wood shall we take, as Frost asked, knowing that stepping down it will make all the difference?

Green Voices in Concert

In the end, our society will be defined not only by what we create, but by what we refuse to destroy.
—John C. Sawhill

Given time—time not in years but in millennia—life adjusts, and a balance has been reached. For time is the essential ingredient; but in the modern world there is no time . . . the chemicals to which life is asked to make its adjustment are no longer merely the calcium and silica and copper and all the rest of the minerals washed out of the rocks and carried in rivers to the sea; they are the synthetic creations of man's inventive mind, brewed in his laboratories, and having no counterparts in nature.
—Rachel Carson, *Silent Spring*

The past grows longer, and the future grows shorter.
—White Feather

When we destroy the earth and everything on it, we are destroying the future of our children and grandchildren. Economic earnings are good, but it's better to leave as legacy a natural wealth that can create new life.
—Jesus Leon Santos, *Treehugger.com*

All things new must change and only that which changes remains true.
—Carl G. Jung

He [a human] commonly thinks of himself as having been here since the beginning—older than the crab—and he also likes to think he's destined to stay to the bitter end. Actually, he's a latecomer and there are moments when he shows every sign of being an early leaver, a patron who bows out after a few gaudy and memorable scenes.
—E. B. White, *Second Tree from the Corner*

We've known for some time that we have to worry about the impacts of climate change on our children's and grandchildren's generations. But now we have to worry about ourselves as well.
—Margaret Beckett

We are in trouble just now because we do not have a good story. We are in between stories. The old story, the account of how we fit into it, is no longer effective. Yet we have not learned the new story.
—Thomas Berry, *Dream of the Earth*

By loving the natural environment as we would a child, the environment will love and serve our children, producing food and clean water for future generations.
—Feliciano dos Santos

We must create an environment in which change enlivens and enriches the individual, but does not overwhelm him.
—Alvin Toffler, *Future Shock*

What does it mean to be an elder in this culture? What are my new responsibilities? What has to be let go to make room for the transformations of energy that are ready to pour through the body-soul? I don't want to be here if I can't carry my own weight. . . . Do I have the courage to live with this evolving me?
—Marion Woodman, *Bone*

Elderwork is service beyond ego in support of the well-being and viability of the whole. . . . Elderwork implies dropping the personal judgments about whether your experiences were "good" or "bad" and seeing that all that has happened in your life has contributed to the creation of your wisdom. . . . Elders are not to be confused with charismatic leaders or heroes; they can be highly visible or go completely unnoticed. . . . An elder may or may not be old chronologically. . . . No one is too young or too busy to serve.

—Barbara Shipka, *Leadership in a Challenging World*

People take the long view when they feel a commitment to those who come after them.

—Rosabeth Moss Kanter

There are limits to growth. The Earth is a closed system and it can support only a finite number of human beings. . . . Mountains of scientific evidence suggest that some limits have already been crossed. Given these conditions, growth cannot continue much longer.

There are no limits to development. The way we live can always be made better; more beautiful, more inventive, more creative, more efficient, more fulfilling. . . . Since there is no limit on humanity's capacity to evolve, development can go on virtually forever.

—Alan AtKisson, *Believing Cassandra*

Like any revolution in thought on a national scale, this one will be a bottom-up phenomenon.

—Robert L. Nadeau, *The Environmental Endgame*

There is great danger in the perception that "there are too many people," whatever truth may be in it, for this is a premise from which it is too likely that somebody, sooner or later, will proceed to a determination of who *are the surplus. . . . Which ones, now apparently unnecessary, may turn out later to be indispensable. We do not know; it is a part of our mystery, our wildness, that we do not know.*

—Wendell Berry, *Preserving Wildness*

A sustainable system is nature's version of the proverb, "waste not, want not."

—Gary Hirshberg, *Stirring it Up: How to Make Money and Save the World*

When we see the rich nations become richer and the poor nations become poorer while they grow in population, the number of time bombs that are planted round us—radioactivity, overpopulation, destruction of nonrenewable resources—is such that we're inclined to yell, "Stop it!" We have to do something, we have to put tremendous pressure on our governments to stop these things. Our indignation must be told. . . .
We have to yell it.
—Jacques Cousteau

A sustainable world is not an impoverished world, but one that is prosperous in different ways. The challenge for the 21st century is to create that world.
—Tim Jackson, Worldwatch.org

The end of inflationary growth on Earth may cause global disruptions—economies collapsing, resource wars, plagues, climate destabilization, mass migrations. But we have an extraordinary opportunity that has arisen only twice before in the history of Western civilization—the opportunity to see everything afresh through a new cosmological lens. . . . But we'll need to change.

—Joel R. Primack and Nancy Ellen Abrams, *The View from the Center of the Universe*

Most human beings are not couch potatoes and whiners. We are doers and creators. In fact, humans need to "make a dent."

—Frances Moore Lappé, *YES!* Magazine

A society headed for overshoot and collapse is therefore, by definition, unsustainable.
—Alan AtKisson, *Believing Cassandra*

A sustainable culture recognizes relationships. That is, it knows that everything is connected. . . . Sustainability is not a state we reach but something we work toward forever.
—Gary Holthaus, *Learning Native Wisdom*

Why does resiliency surprise us? We are born survivors.
—Diane Ackerman, *An Alchemy of Mind*

We lived for millennia as children of the Earth before we built our cities and those modern industrial "inner city areas" whose severance from the natural world is an outrage to the souls of their inhabitants, whether they know it or not. To be severed from nature and from all those images that remind us of who we are causes sorrow—depression it is called nowadays—and violence.
—Kathleen Raine, "Disconnected Souls," quoted in *Cherish the Earth*, ed. Mary Low

And God showed me a little thing, in the palm of my hand, round like a ball, no bigger than a hazelnut. I gazed at it, puzzling at what it might be. And God said to me,

"It is all of creation. . . . It lasts, both now and forever, because I cherish it."
—Julian of Norwich, *Showings*

When despair for the world grows in me,
And I awake in the night at the least sound
In fear of what my life and my children's lives may be,
I go and lie down where the wood drake
rests in his beauty on the water, and the great heron
feeds.
I come into the peace of wild things
Who do not tax their lives with forethought
Of grief. I come into the presence of still water.
—Wendell Berry, "The Peace of Wild Things"

When we unblock our despair, everything else follows—the respect and awe, the love.
—John Seed

To live is to wrestle with despair yet never to allow despair to have the last word.
—Cornel West, Restoring Hope

May the Star of Truth call forth a web of protection.
May truth be set free
May truth be brought from the shadows into light.
May the innocent be protected and set free
May truth-speakers be protected and empowered.
—Anne Dosher

It is not half so important to know as to feel when introducing a young child to the natural world.
—Rachel Carson

Becoming a larger Self that knows its capacity,
Together, we create a future worth living.
—Nancy Marguilies, Juanita Brown, *The World Café*

It will require that we re-imagine our lives. . . . It will require of many of us a humanity we've not yet mustered, and a grace we were not aware we desired until we had tasted it.
—Barry Lopez

Throughout history, the really fundamental changes in societies have come about not from dictates of governments and the results of battles but through a vast number of people changing their minds—sometimes only a little bit.
—Willis Harman, *Global Mind Change*

Why aren't more people getting involved? For some it's simply a lack of understanding. . . . Once individuals grasp the scope of the problem, they often resort to denial. Political leaders reinforce this resistance to change, proposing remedies that skirt the real problems at hand.

—Betsy Taylor, *Sustainable Planet*

We are all sailors on an unknown sea; may He make us brave enough to accept this mystery.

—Paul Coelho, *Brida*

But here's the good news. Just as the problems are interconnected, so too are the solutions. Solving one part of the problem can have a positive ripple effect.
—Shannon Daley-Harris and Jeffrey Keenan with Karen Speerstra, *Our Day to End Poverty*

Awakened doubt can nudge us toward the creative experimentation that can result in our changing our minds in the small ways that eventually add up to major change.
—Carol Frenier, *Business and the Feminine Principle*

For civilization as a whole, the faith that is so essential to restore the balance now missing in our relationship to the earth is the faith that we do have a future. We can believe in that future and work to achieve it and preserve it, or we can whirl blindly on, behaving as if one day there will be no children to inherit our legacy. The choice is ours; the earth is in the balance.
—Al Gore, *Earth in the Balance*

We are in the early phase of a revolution that will fundamentally transform enterprises around the planet. The transformation is a paradigm shift of immense magnitude.
—James Martin, *The Great Transition*

Almost anything you do will seem insignificant but it is very important that you do it. . . . You must be the change you wish to see in the world.
—Mahatma Gandhi

The heroes of all time have gone before us . . . and where we had thought to find an abomination, we shall find a god; where we had thought to travel outward, we shall come to the center of our own existence; where we had thought to be alone, we shall be with all the world.
—Joseph Campbell

To live more voluntarily is to live more deliberately, intentionally and purposefully—in short, it is to live more consciously. We cannot be deliberate when we are distracted from life. We cannot be intentional when we are not paying attention. We cannot be purposeful when we are not being present. Therefore, to act in a voluntary manner is to be aware of ourselves as we move through life.
—Duane Elgin, *Voluntary Simplicity*

Politics, not droughts or overpopulation, cause hunger.
—Dennis Avery

Stewardship companies understand, whether consciously or instinctively, the power of these biophilic urges for connection and serving the quality of life. . . . Profit can arise only from life [and] in a healthy world, profit must serve life. . . . Like deep ecology, deep stewardship requires a fundamental shift in the way we see ourselves in relation to the rest of life. . . . There is no time to waste in shifting to the living company/living assets paradigm. . . . We must abandon the traditional machine model of the firm and its destructive ways, and move with speed and intention to a deep stewardship model that respects life. We have the capacity to do so. And we need to do it now.

—Joseph H. Bragdon, *Profit for Life*

Change can come fast once the times are right.

—John Passacantando, Greenpeace

Others have said we are deep into a Great Forgetting, a time when our relationships are so fractured that we have almost forgotten why they were important in the first place. I prefer to believe that we are on the brink of a Great Remembering, a time when we can reconsider what matters most to us, when we are punished enough by "the world I know is gone" that we can find the courage to ask ourselves, how much is enough?
—Peter Forbes, "Another Way of Being Human," quoted in *Sustainable Planet*, ed. Juliet B. Schor and Betsy Taylor

I would like to do whatever it is that presses the essence from the hour.
—Mary Oliver

Abundance is a communal act, the joint creation of an incredibly complex ecology in which each part functions on behalf of the whole and, in return, is sustained by the whole. Community not only creates abundance—community is abundance. If we could learn that equation from the world of nature, the human world might be transformed.

—Parker Palmer, "There is a Season," quoted in *The Impossible Will Take a Little While*, ed. Paul Rogat Loeb

An open bank account allows money to flow in. An open heart allows money to flow out. In the end, it's flowing love that counts.

—Meredith Young-Sowers, *Wisdom Bowls*

Real wealth is found among people who have a sound sense of their place in the world, who link their actions and thoughts with those of others and who are strong, vigorous and co-operative actors in their communities and ecosystems. Rich are those people who balance the benefits they receive in life with the responsibilities they assume for themselves, their families and communities and their environment.
—Nancy J. Turner, *The Earth's Blanket*

A machine cosmology gives rise to unemployment and scarcity, where a cosmology of connection would encourage economics of abundance.
—Matthew Fox

Most people in North America engage in the dominant myth of "more is better" without question, and even good, caring people rationalize excess as necessity. . . . "More is better" now means "more money is better." This acquiescence to excess then requires putting up thicker and thicker walls between our consciences and the billions of people who live in poverty.
—Vicki Robin, *Your Money or Your Life*

The future of the world hangs on the breath of schoolchildren.
—Hebrew proverb

The worldview is based on scarcity, which in turn creates scarcity and conflict and poverty. It cannot create abundance. A worldview of abundance is the worldview of women in India who grow food not just for humans, but for all species. They leave food for ants on their doorstep, even as they create the most beautiful art in kolams, mandalas, and rangoli with rice flour. This view of abundance recognizes that, in giving food to other beings and species, we maintain conditions for our own food security. In feeding the earthworms, we feed ourselves. In feeding cows, we feed the soil, and in providing food for the soil, we provide food for humans. This worldview of abundance is based on sharing and on a deep awareness of humans as members of the earth family.

—Vandana Shiva, speech, 2000

I love to play indoors better 'cause that's where all the electrical outlets are.
—A San Diego fourth-grader, quoted in Richard Louv, *Last Child in the Woods*

If children do not attach to the land, they will not reap the psychological and spiritual benefits they can glean from nature, nor will they feel a long-term commitment to the environment, to the place.
—Richard Louv, *Last Child in the Woods*

Children should be out here on the water. This is what connects us, this is what connects humanity, this is what we have in common. It's not the Internet, it's the oceans.

—Robert Kennedy, Jr.

The street, the square, the park, the market, the playground are the river of life. We are now rediscovering how essential they are to our well-being.

—Kathy Madden, quoted in Jay Walljasper, *The Great Neighborhood Book*

Grasping the Grand Scheme is demanding for adults; for kids raised on Disney, it's simply shocking to discover that it takes a bunch of Bambis to feed a Lion King, and that Mowgli's wolves would eat Thumper and all his sibs. Eventually, most of us figure out that it's people, not nature, who create morality, values, ethics—and even the idea that nature itself is something worth preserving. We choose to be shepherds and stewards, or we don't. We will live wisely—preserving water and air and everything else intrinsic to the equations we're only beginning to understand—or we won't, in which case Nature will fill the vacuum we leave. She is exquisite, and utterly indifferent.

—Seth Norman, quoted in Richard Louv, *Last Child in the Woods*

I'm only a child, yet I know if all the money spent on war was spent on ending poverty and finding environmental answers, what a wonderful place this Earth would be . . . You grownups say you love us, but I challenge you. Please make your actions reflect your words.
—Severn Cullis-Suzuki, age 12, quoted in *Sacred Balance*, David Suzuki

The prison is your culture, which you sustain generation after generation. . . . When six billion of you refuse to teach your children how to be prisoners of Taker Culture, this awful dream of yours will be over—in a single generation. It can only continue for as long as you perpetuate it. . . . If you cease to perpetuate it, then it will vanish.
—Daniel Quinn, *My Ishmael*

We have become a nation that places a lower priority on teaching its children how to thrive socially, intellectually, even spiritually, than it does on training them to consume. The long-term consequences of this development are ominous.
—Juliet B. Schor, *Born to Buy*

If small children are treated with respect and the issues are explained to them in a simple but powerful manner, they are great lobbyists. Who is better prepared to convey a strong message defending the future of our natural resources than our children?
—Rosa Hilda Ramos

We need to take care of mother earth because we need strength, life and beauty.
—Sherlinda Nahwahquaw (Desert Sky), MITW Tribal school fourth-grader

In twenty years, our globe will be trying to support 8 billion people, one third of them children—children who are wondering how we adults could have messed things up so badly. Children who want to live safe, clean, secure lives. Children who look around and wonder what will be left for them. Children who will have talents and the intelligence to create different ways of living, but will wonder why it took us so long to realize what was happening to the Earth.
—Alex Steffen, *TED.com*

One way to help Mother earth is to clean up all the junk everywhere and not to waste anything. Try to recycle as much as you can. . . . I think they should have a machine that makes garbage smaller instead of just piling it up. But it's better if there's less garbage. If you think really hard, you can find things to make out of things that are broken.

—Naneque Latendeer (Happy Woman), MITW Tribal School fourth-grader

Teaching children about the natural world should be treated as one of the most important events in their lives.

—Thomas Berry

It's vitally important for our children to know that the curse that needs to be lifted from the earth is not humanity. It's important for them to know that we may be a doomed culture, *but we are not a doomed* species. *It's important for them to understand that it's not being HUMAN that is destroying the world. It's living THIS WAY that is destroying the world. It's important for them to know that humans HAVE lived other ways, because it's important for them to know that it's POSSIBLE for humans to live other ways. Otherwise they can only repeat the falsehood spoken by that waste disposal engineer: That the only way to stop poisoning the world is to get rid of humanity.*
—Daniel Quinn, speech, 1997

Believe we can save all of our children and then do it.
—Marion Edelman

In general, I feel optimistic about the future. The rapid changes in our attitude toward the earth are also a source of hope. As recently as a decade ago, we thoughtlessly devoured the resources of the world, as if there were no end to them. We failed to realize that unchecked consumerism was disastrous for both the environment and social welfare. Now, individuals and governments are seeking a new ecological and economic order.
—The Dalai Lama, speech, 1992

Good intentions are great, but remember that only passion changes the world.
—Alex Steffen, Worldchanging: *A User's Guide for the 21st Century*

The Great Work now . . . is to carry out the transition from a period of human destruction of the Earth to a period when humans would be present to the planet in a mutually beneficial manner. . . . So now we awaken . . . .

—Thomas Berry, *The Great Work*

I believe that we are on the verge of a paradigm shift, and what the leaders in that new paradigm do and say—the ones who are invisible on the cutting edge today—will soon be seen as mainstream. . . . But finding a meaningful myth, a "story of what it is all about," that would fit a modern society, is not easy. . . . Cultures must grow organically—when the time is right for it to happen, and when people want it.

—Karl-Henrik Robèrt, speech, 2001

Ecological sensibility will not follow man-made clock time. It will experience time as flux and flow with its stream. . . . It will develop an intimate relationship with other animals and plants. . . . Ecological sensibility will become aware of what one already is and open up to what one already has.
—Peter Marshall, *Nature's Web*

To be effective, conservation must be guided by a vision that is bold, scientifically credible, practically achievable, and hopeful.
—Dave Foreman, The Rewilding Institute

We ALL have to make a difference. It doesn't matter what job we do. We can't have people saying, "Oh, I just flip burgers, so I can't make a difference." "Oh, I just drive a cab, so I can't make a difference." "Oh, I just sell insurance, so I can't make a difference." "Oh, I'm just an auto mechanic, so I can't make a difference." "Oh, I'm just an accountant, so I can't make a difference." Concentrate on doing what you do best, because THAT'S where you'll have the most influence on the future of the world.

—Daniel Quinn, speech, 1997

Search for life, understand it and above all, preserve it.

—E. O. Wilson

Communities [are] solving environmental problems using the power of harnessed information. [This wave of environmentalism] will weave together the grass roots and the nationals through technology. There will be more bottom-up empowerment and a tremendous trend toward pluralism in the movement because there has to be if we are to solve our problems.
—Fred Krupp, quoted in *Earth Rising*, Philip Shabecoff

Listen to the course of being in the world . . . and bring it to reality as it desires.
—Martin Buber

If global civilization can self-organize from our current chaos, it will be founded on cooperation rather than winner-take-all competition, sufficiency rather than surfeit, communal solidarity rather than individual elitism, reasserting the sacred nature of all earthly life. Those who desire such a world will work to create it.
—Daniel Pinchbeck, *2012: The Return of Quetzalcoatl*

All things are possible once enough human beings realize that everything is at stake.
—Norman Cousins

After the final no there comes a yes
And on that yes, the future world depends.
—Wallace Stevens

Here in the Appalachians, on some not-too-distant day, I will wake up and drink a glass of fresh milk from a neighbor's small dairy; that night I will hear a pack of wolves, howling from Buck Hill. And it will raise the hair on my arms, and it will fill me up with hope. Hope that the greenhouse effect might someday abate. Hope that this society might be starting the climb down from overdevelopment. Not hope that everything will be fine—everything isn't going to be fine. But hope that the sky is brightening a little in the East.
—Bill McKibben, *Hope, Human and Wild*

How will we save the planet? Only by awakening to who we are. And how will we act when we awaken to who we are?
—Andrew Harvey, *The Way of Passion: A Celebration of Rumi*

At this time in history, we are to take nothing personally—least of all ourselves. For the moment that we do, our spiritual growth and journey comes to a halt. The time of the lone wolf is over. Gather yourselves; banish the word "struggle" from your attitude and your vocabulary. All that we do now must be done in a sacred manner and in celebration. We are the ones we've been waiting for!
—Hopi elder

Go forth on your journey,
for the benefit of the many,
for the joy of the many,
out of compassion for the welfare,
the benefit and joy of all beings.
—The Buddha

This is not the time to stay home,
But to go out and give yourself to the garden . . .
It is the time of grace, it is the time of generosity . . .
Dance in the new world.
—Rumi

6

Space

Our tiny blue and white pearl floats in the black velvet universe, fifth largest of the nine known planets, third from the sun. Greek stories tell how, long before humans came to earth, the round disk of earth consisted of the Friendly Sea and the Unfriendly Sea and around the earth a great river, Ocean.

Far to the north lived the Hyperboreans, along with the Muses with lyres and flutes. The beings who populated this frosty garden knew no sickness, no old age, no death. Zeus decided, one day, it was time for animals to arrive on earth. He gave them strength and courage, cunning and swiftness. Then two Titans, Epimetheus (Afterthought) and his brother Prometheus (Forethought), created humans. In order to give humans a fighting chance on earth, Prometheus traveled to the sun and brought back fire in a fennel tube: the gift of sustainability. The word *sustainable* comes from the Latin, meaning "to hold up," "to continue indefinitely." In other words, it represents something that can be maintained over time. But what we have learned is that with or without Prometheus's gift, unsustainable societies collapse.

Remember when archeologists unearthed "Lucy," a tiny skeleton estimated to have lived 3.2 million years ago? Experts posit she would have weighed about sixty-five pounds and measured roughly four feet tall. The workmen in the Afar badlands of Hadar in north-central Ethiopia who found her named her *Denkenesh*—"You are wonderful."

That little skeleton, and everything currently living, makes up the "The Great Clod Project," as the poet Gary Snyder put it. He worked on his eco-cultural history of China by that name in the 1970s, and then later used the title for a published collection of poems. In his introduction, he wrote: "Each living being is a swirl in the flow, a formal turbulence, a song. The land, the planet itself, is also a living being—at another pace." We swirl, flow, and leave tracks for those who will come after.

We know we're all connected. At a deep level, we are upset when whales are killed or dolphins beach; we weep at pictures of fatigued polar bears stranded on floating ice or swimming across vast stretches of water that used to be ice. Why? *Because they are us.* In nature there really is no "other," no matter how different we may think we are from one another.

We remind ourselves of this whenever we use the word *Namaste*—the God in me sees the God in you. Or the Mayan phrase *Lak'esh*—you are another me. What I do to you, I do to myself.

Because everything is connected, when we pull a thread out of this intricate tapestry of life, there are always unimagined consequences. Things fray and unravel. Bats develop fungus. Bees falter. Life-giving plants disappear. Then we humans, in turn, get sick.

Animals and insects seem to understand our connectedness, but most of us humans are just beginning to remember. Certain birds, for instance, weave medicinal plants into their nests to keep pests from harming their young. Dogs will eat quack grass for health reasons. Medicinal herbs have been found in 40,000-year-old Neanderthal graves. Female elephants use the same plant to bring on childbirth that Kenyan women have used for centuries. When a female elephant's pregnancy is nearing its end, she will sometimes travel up to twenty-eight miles to find that plant and eat it, for the sake of her child.

The phrase *deep ecology* was introduced in the late 1960s, at a time when caring for the environment was foolishly politicized. Like Denkenesh, the phrase means that *all* life is sacred and valuable—all individuals, regardless of species, have the right to live. It brings with it the understanding that we are here to embrace and support the world—not control, use, or conquer it.

We once shot passenger pigeons for their tongues. We killed buffalos from the parlor cars of trains for target practice. That disposition is still alive and well. People who belong to the "Red Mist Society" gather to shoot prairie dogs in the West—the earth misted with the blood of prairie dogs who struggle to support over two hundred other species of living beings. Weavers use the hides of Tibetan antelope to make six-foot-long shawls. A single Kashmiri shawl made from this "wool of kings" can cost upwards of $17,000. Movie stars and royals wear them. About 20,000 antelopes are killed each year in this pursuit of very high fashion. There may be only 75,000 Tibetan antelope left.

Extinction. The word tastes bad on the tongue. Maybe because it sounds like *stink.* It's a flat, dark word, a final word—as opposed to a living, vibrant word like *green.* Yes, animals do disappear over time. All large Pleistocene animals, for instance, are now extinct except the musk ox and the reindeer. But animals don't usually disappear without some help from humans. The last Yangtze River freshwater dolphin stopped swimming in 2007. In the 1950s, thousands of these eight-foot-long mammals, some weighing a quarter of a ton, playfully swam up and down the Yangtze, but that was before dams stopped the flow. The last two giant soft-shelled turtles may still be living in a Chinese zoo. Or not. Where is Noah now? And who among us is able to hammer a newfangled ark?

A yellow, slimy enzyme nestles below the boardwalks surrounding Yellowstone's hot springs. This heat-resistant enzyme is crucial for what scientists call a polymerase chain reaction. If it disappears, so will your doctor's ability to use it to quickly swab, get a culture, and diagnose your sore throat. We live suspended on a delicate chain of life. By destroying *any* of the links, we weaken our planet and ourselves.

Biologist E. O. Wilson coined the acronym HIPPO to help us remember how we are deliberately destroying our planet: H, habitat destruction; I, invasive species; P, pollution; P, population increase; and O, over-harvesting. Wilson suggests that by promoting biodiversity, we can protect Creation itself.

As the poet Adrienne Rich put it: "My heart is moved by all I cannot save." Our hearts *are* moved, and we wonder what we can do. We've changed our lightbulbs; we carry canvas bags to the supermarket—when we remember to put them in our car. But we still pump 85 million barrels of petroleum a day out of the ground. And then we burn it up. From dawn to dusk, we send waste from 27 billion pounds of coal up into our precious air. Every day! In the Amazon alone, five million acres of rain forest disappear each year. Why? Because consumers (that's us) around the world demand hardwoods, soybeans, beef, coffee. So trees are cut to accommodate those needs, offering livelihood for the people who once sustainably lived in those forests.

Paul Hawken, a longtime writer on all things ecological, estimates in his book, *Blessed Unrest,* that there are now about *a million* organizations working toward ecological sustainability and social justice. It's not a movement in a conventional sense because there are no members. It doesn't have one doctrine; it's fiercely independent and dispersed around the globe. The poet Gary Snyder calls it "the great underground" and claims it's been a current of our humanity way back to the Paleolithic time. This movement has three basic roots: environmental activism, social justice initiatives, and wisdom from indigenous cultures. These three roots now emerge as one green voice.

The Native Americans who lived on the west coast of Canada and southeastern Alaska called each cedar tree "Life Giver." They harvested and used every part. Some cherished cedar trunks became four-story-tall totem poles notched with carefully carved clan stories. Today, native carvers still boldly honor the swimmers, the winged, the two- and four-legged ones. Everything contributes to life. All is *Denkenesh.* Wonderful!

Green Voices in Concert

If you kill off the prairie dogs, there will be no one to cry for rain.
—Navajo warning

Our little Spaceship Earth is only eight thousand miles in diameter . . . so extraordinarily well invented and designed that to our knowledge humans have been on board it for two million years, not even knowing that they were on board a ship.
—R. Buckminster Fuller, *Operating Manual for Spaceship Earth*

It's so incredibly impressive when you look back at our planet from out there in space and you realize so forcibly that it's a closed system—that we don't have any unlimited resources, that there's only so much air and so much water. You get out there in space and you say to yourself, "That's home. That's the only home we have." The first few times we circled the earth, we saw the United States, Europe, Asia, Africa, and the Soviet Union. But as we moved further into space, those imaginary lines began to disappear and we saw that we were one. This profoundly affected many of us. Amidst all the beauty was the profound realization that all was not right with the earth.

—Edgar Mitchell

Viewed from the distance of the moon, the astonishing thing about the earth, catching the breath, is that it is alive. . . . If you had been looking for a very long, geologic time, you could have seen the continents themselves in motion, drifting apart on their crustal plates, held afloat by the fire beneath. It has the organized, self-contained look of a live creature, full of information, marvelously skilled in handling the sun.
—Lewis Thomas, *The Lives of a Cell*

To become aware of other creatures as individuals is to discover that life is a mansion with many rooms.
—Lorraine Anderson, *Sisters of the Earth*

We must convince each generation that they are transient passengers on this planet earth. It does not belong to them. They are not free to doom generations yet unborn. They are not at liberty to erase humanity's past nor dim its future.
—Bernard Lown and Evjueni Chazov, quoted in *Peace: A Dream Unfolding*

This star, our own good earth, made many a successful journey around the heavens ere man was made, and whole kingdoms of creatures enjoyed existence and returned to dust ere man appeared to claim them. After human beings have also played their part in Creation's plan, they too may disappear without any general burning or extraordinary commotion whatever.
—John Muir, *A Thousand-Mile Walk to the Gulf*

Creation in all its ravishing beauty, with its infinite baroque embellishments and subtle charms, with all the wonders that it offers from both the Maker and the Made, with all its velvet mystery and with all the joy we receive from those we love here, so enchants us that we lack the imagination, less than the faith, to envision an even more dazzling world beyond, and therefore even if we believe, we cling tenaciously to this existence, to sweet familiarity, fearful that all conceivable paradises will prove wanting by comparison.
—Dean Koontz, *One Door Away from Heaven*

Modern man, the world eater, respects no space and no thing green or furred as sacred. The march of the machines has entered his blood.
—Loren Eiseley, *The Invisible Pyramid*

To play Earth Jazz, you must first of all listen.
—Evan Eisenberg, *The Ecology of Eden*

We experience ourselves, our thoughts and feelings, as something separated from the rest—a kind of optical delusion of consciousness. The delusion is a kind of prison for us, restricting us to our personal desires and to affection for a few persons nearest to us. Our task must be to free ourselves from this prison by widening our circles of compassion to embrace all living creatures and the whole of nature in its beauty.
—Albert Einstein, quoted in *Science, Soul and the Spirit of Nature*, Irene van Lippe-Biesterfield and Jessica van Tijn

What we most need to do is to hear within us the sounds of the Earth crying.
—Thich Nhat Hanh

I think it was from the animals
that St. Francis learned
It is possible to cast yourself
On the earth's good mercy and live.
—Jane Hirschfield, quoted in *The Re-Enchantment of Everyday Life*, Thomas Moore

The hour is late and the stakes are higher than ever, but we have no other course than to learn how to intervene more responsibly to heal the Earth.
—Hazel Henderson

God arranged the order of creation so that all things are bound to each other. The direction of events in the lower world depends on entities above them, as our sages teach: "There is no blade of grass in the world below that does not have an angel over it striking it and telling it to grow."
—Aryeh Kaplan, *Meditation and Kabbala*

The partnership between the bear and the Pawnee, or between the owl and the Kwakiutl, is such that each shares the fate of the other.
—John Bierhorst, *The Way of the Earth*

A land without wildflowers will be a hotel, not a homeland.
—Yizhar Smilansky

Ignoring the evolutionary attributes of biological systems can only be done at the peril of ecological catastrophe.
—Marc Lappé

We are Nature, long have we been absent, but now we will return.
—Walt Whitman

Man's heart, away from nature, becomes hard; [the Lakota] knew that lack of respect for growing, living things soon led to lack of respect for humans too.
—Luther Standing Bear

[In ancient China] a painting was not exhibited, but unfurled before an art lover in a fitting state of grace; its function was to deepen and enhance his communion with nature.
—Lawrence Leshan and Henry Margenau, *Einstein's Space and Van Gogh's Sky*

Nothing is itself without everything else.
—Thomas Berry

If we had a keen vision and feeling of all ordinary human life, it would be like hearing the grass grow and the squirrel's heart beat, and we should die of that roar which lies on the other side of silence. As it is, the quickest of us walk about well-wadded with stupidity.
—George Eliot, Middlemarch

A significant part of the pleasure of eating is in one's accurate consciousness of the lives and the world from which the food comes.
—Wendell Berry, *What Are People For?*

In Japan in the spring we eat cucumbers.
—Shunryu Suzuki

If you kill the crows, you'd *have to take care of the roadkill yourselves. Let crows be crows.*
—Daniel Quinn, *My Ishmael*

Destroying a tropical rainforest and other species-rich ecosystems for profit is like burning all the paintings of the Louvre to cook dinner.
—E. O. Wilson, quoted in *Hot, Flat and Crowded*, Thomas L. Friedman

Live in each season as it passes; breathe the air, drink the drink, taste the fruit, and resign yourself to the influences of each.
—Henry David Thoreau, *Huckleberries*

When your mother told you to eat everything on your plate because people were starving in India, you thought it was pretty silly. You knew that the family dog would be the only one affected by what you did or didn't waste. . . . The act of putting into your mouth what the earth has grown is perhaps your most direct interaction with the earth.
—Frances Moore Lappé, *Diet for a Small Planet*

Earth, we might say, is a single reality composed of a diversity beyond all understanding or description.
—Thomas Berry, *The Great Work*

The world's food supply hangs by a slender thread of biodiversity.
—E. O. Wilson, *The Future of Life*

The biodiversity of the planet is a unique and uniquely valuable library that we have been steadily burning down—one wing at a time—before we have even cataloged all the books, let alone read them all.
—John Holdren

From genes to organisms to ecosystems to cultures—at every level the patchwork diversity adds up to a single living whole.
—David Suzuki, *The Sacred Balance*

A sustainable culture also takes care of ethnic diversity, including language diversity, ceremonial and ritual diversity, and diversity of worldview, recognizing that these diversities are as critical as biological diversity to the world's survival.
—Gary Holthaus, *Learning Native Wisdom*

In ecosystems, members seem to have access to the whole system. The quality of their communication is dazzling. Birds build their nests over a river at different heights each year in anticipation of the coming flood levels. Furry animals know how much snow to expect in the coming winter and dress accordingly.
—Margaret J. Wheatley and Myron Kellner-Rogers, *A Simpler Way*

An entire ecosystem can exist in the plumage of a bird.
—E. O. Wilson

The ecological model is a model of internal relations. No event first occurs and then relates to the world. The event is a synthesis of relations to other events . . . the elements in the cell relate to one another and to the cell as a whole more like the way the animal as a whole relates to its environment.
—Charles Birch

Once something has been lost, it's lost. And once that something—either a brightly colored bird or miles of undeveloped land—is gone, it becomes impossible for us to understand the positive ways in which it might have affected us.
—Emily White, "Greening the Blues"

If we were to wipe out insects on this planet, which we are trying hard to do, the rest of life and humanity with it would mostly disappear from the land, and within a few months.
—E. O. Wilson, *TED.com*

Maintaining a population of six billion humans costs the world two hundred species a day. If this were something that was going to stop next week or next month, that would be okay. But the unfortunate fact is that it's not. It's something that's going to go on happening every day, day after day after day—and that's what makes it unsustainable, by definition. *That kind of cataclysmic destruction* cannot *be sustained.*
—Daniel Quinn

I think that, in the long run, in order to participate joyfully and wholeheartedly in the deep ecology movement, you have to take your own life very seriously.
—Arne Naess, *The Ten Directions*

Our thoughts must be on how to restore to the Earth its dignity as a living being, how to stop raping and plundering it as a matter of course. We must begin to develop the consciousness that everything has equal rights because existence itself is equal. In other words, we are all here: trees, people, snakes alike. We must realize that even tiny insects in the South American jungle know how to make plastic, for instance; they have simply chosen not to cover the earth with it.
—Alice Walker, *Everything Is a Human Being*

The library of life is burning and we do not even know the titles of the books.
—Gro Harlem Brundtland

Once a species is gone, it is lost forever—we have lost that million years of our planetary heritage forever.
—Thomas Brooks

Animals, plants, and insects do not have to become extinct for a ecosystem to begin a wobble toward chaos.
—Eugene Linden, *The Future in Plain Sight*

On the road to extinction, traffic travels both ways.
—Kenneth Brower

The world is no longer divided by the ideologies of left and right, but by those who accept ecological limits and those who don't.
—Wolfgang Sachs

The way to an ecological way of life is to treat our houses as homes, our communities as homes, and nature as home. It is the intimacy in each relationship that serves the welfare of the other; at root, ecology is an erotic attitude of closeness, relatedness, and care.
—Thomas Moore, *The Re-Enchantment of Everyday Life*

Whenever we diminish biological diversity, we are always more vulnerable.

We are plain citizens of the larger biotic community. . . . Can the wave and the ocean be separated? Or the tree and the leaf?

—Aldo Leopold, *Sand County Almanac*

Small is not necessarily beautiful (consider the starry heaven or the wide ocean), but eco-communities should be shaped on a human scale in which people can enjoy face-to-face relationships. . . . We come from nothing. It is too early for us to return to nothing.

—Peter Marshall, *Nature's Web*

Deep ecology recognizes that nothing short of a total revolution in consciousness will be of lasting use in preserving the life-support systems of our planet.
—John Seed, *Thinking Like a Mountain*

What we must do is incorporate the other people . . . the creeping people, and the standing people, and the flying people and the swimming people . . . into the councils of government.
—Gary Snyder, quoted in *The Man Who Talks to Whales*, Jim Nollman

It's all one seal.
—Eskimo saying

When I was at school my jography told as th'world was shaped like a orange an' I found out before I was ten that th'whole orange doesn't belong to nobody. No one owns more than his bit of a quarter an' there's times it seems like there's not eow quarters to go round. But don't you—none o'you—think as you own th' whole orange or you'll find out you're mistaken, an' you won't find it out without hard knocks . . . there's no sense in grabbin' at th'whole orange—peel an' all. If you do you'll likely not get even th' pips, an' them's too bitter to eat. (Susan Sowerby)
—Frances Hodgson Burnett, *The Secret Garden*

7

Essence

Medieval philosophers thought *Quintessence* was the fifth and highest element in existence. It permeates all of nature and reaches out to form the stars. It's the pure essence of everything: Soul. Light. Ether. It's what passes through our bodies, according to Hildegard of Bingen, like sap through a tree. The soul sustains the body, and in turn the body sustains the soul. Each of us senses this burning desire to experience something more.

Thomas Moore, former monk, professor of religion and psychology, and author of many books, including *Original Self*, says that the soul doesn't evolve or grow, but cycles and repeats: "echoing ancient themes common to all human beings. It is always circling home."

Rudolph Steiner, in *Theosophy*, calls the soul the keeper of the past that continually collects treasures for the spirit.

As we find ourselves nearing the end of this book, we sense a circling homeness, a quickening spirituality that permeates ourselves and our planet. We collect, just by living, treasures for our collective spirits; we expand our communal essence. And by sharing these insights, we continue to expand the "green devotional" ever outward.

In the Green Mountains of Vermont, several women live their essence in plain sight. They refer to their community of prayer and contemplative labor as *Ecozoic:* House of Life. Gail Worcelo and Bernadette Bostwick, under the guidance of Thomas Berry and inspired by the Rule of St. Benedict, are engaged, with other sisters, in what they call "green monasticism." Old symbols take on new meanings for them. For instance, they have created a litany about rocks. In their "A Celebration in Stone," they name and touch the ancient ones: granite, marble, slate, quartzite, shale, schist, gneiss, greenstone, serpentine, limestone. Each touch takes them back in time, and they experience awe in the presence of a council of "elders" that form our Mother Earth. They make clay earth prayer beads and paint icons. Their rule of life is shaped by an understanding of an unfolding universe. They cultivate a community that is ideally a "seamless garment" of words, prayer, and work.

While we may not find ourselves in a reinvented monastery such as the one in Greensboro, each of us can emulate the sisters' ideal of remembering our essence in our words, our prayers, and our daily work. We will be less stressed, less fearful, and less depressed about the state our earth.

Still, knowing what we know, we may have insomnia from time to time. And we may lie in bed some mornings just waiting for the dawn.

But when does dawn come? That question was once asked by a rabbi. His students offered varying definitions of the time when night ends and day begins. "No," the rabbi said of each definition. "It's when you look into the face of another human being and have enough light to see yourself."

With enough light, we can see our responsibilities and observe how, together, *we* can save this awesome pearl of a planet for future generations. The light has always been there for us, but it's taken us millions of years to develop the wherewithal to interact with the cosmos. Brian Swimme, in *The Universe Is a Green Dragon,* says, "The universe has poured into you the creative powers necessary for further development. . . . For the unfolding of the universe, your creativity is as essential as the creativity inherent in the fireball. . . . The universe continues to unfold, continues to reveal itself to itself through human awareness."

From her fiery core to her floating ozone membrane, something much bigger than ourselves wraps earth and allows our sun, stars, moon, and other planets to shimmer down on us. It is the Quintessence of Life. We are not alone. *Mazel tov,* literally translated, means "May your planet be favorable." *Mazel tov,* then, to us all!

Green Voices in Concert

The universe flared forth fifteen billion years ago in a trillion-degree blaze of energy, constellated into a hundred billion galaxies, forged the elements deep in the cores of stars, fashioned its matter into living seas, sprouted into advanced organic beings, and spilled over into a form of consciousness that now ponders and shapes the evolutionary Dynamics of Earth.
—Brian Swimme, *Cosmogenesis*

This is holy work. It is fun and it is enlivening. We are co-creators with god—our original occupation. You can't finish the work. This is a teaching from Jewish sources, the Talmud. But you must be part of it. Find your right place in it and get to work! All hands on deck. The only way to get it together is together. This is the soul food we need to fill the void.
—Rabbi Andrea Cohen-Kiener, *Loretto Earth Network News*

Earth is not a what, or a thing—or a god—she is more like something "beyond animal" that is alive, complex, and self-aware in ways that put the broken toys of our common models to shame. . . . Sometime around 3000 million years ago, a unique change took place in the solar family. Allegorically, we might say that an ash-covered princess got a blue dress . . . *3000 million years ago—give or take a passel—Earth began reflecting* blue light.
—Organelle.org

We are all architects of the future. When you have an inspiration, take note. It may only come through once.
—Jacqueline Small

If the word "spiritual" is like the word "whale," then the word "ecology" is like the word "ocean." It is the fluid within which our spirit swims. Spiritual ecology does not place the human race first. It does not place the earth first. In fact, it doesn't place the seventh generation first either. . . . There are no seconds, no thirds, no any other level. The categories drift together.
—Jim Nollman, *Spiritual Ecology*

Man models himself on the Earth;
The Earth models itself on Heaven;
Heaven models itself on the Way;
And the Way models itself on that which is so
on its own.
—Lao Tzu, *Tao Te Ching*

Earth is a germ or seed in the universe. It's growing.
—Rudolph Steiner

We may not be able to turn aside the troubles facing us, but we can cultivate a spiritual life, creating an individual and social spirituality that may short-circuit the violence and lead us to take care of one another and the earth. . . . Difficulty can unite us in a common cause. . . . Remember the Qur'ran's warning? Allah does not change a people's lot unless they change what is in their hearts.
—Gary Holthaus, *Learning Native Wisdom*

Ecology and spirituality are fundamentally connected, because deep ecological awareness, ultimately, is spiritual awareness.
—Fritjof Capra

Vision is not enough. It must be combined with venture. It is not enough to stare up the steps, we must step up the stairs.
—Václav Havel

We will try to probe the mystery of that self-awareness: I am, you are, the universe trying to understand our past so that we can appropriately shape our future. We are the universe beginning to be conscious of itself as the universe.

—Elaine M. Prevallet, *Making the Shift: Seeing Faith Through a New Lens*

Civilized people depend too much on man-made printed pages. I turn to the Great Spirit's book, which is the whole of his creation—study in nature's university...

—Tatanga Mani, *Tatanga Mani*

I am lying in the bosom of the infinite universe, I am at this moment its soul, because I feel all its force and its infinite life as my own.
—Friedrich Schelling, *Romanticism and Revolt*

The rain surrounded the cabin . . . with a whole world of meaning, of secrecy, of rumor. . . . Nobody started it, nobody is going to stop it. It will talk as long as it wants, the rain. As long as it talks I am going to listen.
—Thomas Merton

Whatever we know of the world, there is always more.
—David Bohm, *Changing Consciousness*

Spiritually based sustainable living is an endless dance of reason and faith. Reason without faith succumbs to pride, arrogance, hubris, and all that that brings with it, while faith without reason denies humanity, denies who we are as human beings. Sustainability without attention to Mystery, Spirit, and Spirituality is a dead-end street, for it ignores who we are.
—John E. Carroll, *Sustainability and Spirituality*

I don't think God is going to ask us how he created the earth, but he will ask what we did with what he created.
—Rich Cizik, *New York Times*

Religious environmentalism is a diverse, vibrant, global movement, a rich source of new ideas, institutional commitment, political activism, and spiritual inspiration. . . . Can it make any real difference in the world?

—Roger S. Gottlieb, *A Greener Earth: Religious Environmentalism and Our Planet's Future*

The great curvature of the universe and of the planet Earth must govern the curvature of our own being.

—Thomas Berry

We are hard-wired to care and connect.
—David Korten, *YES!* Magazine

Hymn of the Universe . . . Pray for us
Matter impregnated with Spirit
Fire of the galaxies
Dust of the stars
Music of the spheres
Mother of the Cosmos
Bearer of the silence unfolding
Hidden sense of the ineffable plan
Energy of the supernovas
Space of the spaceless God
Depth beyond imagining . . .
Be our guide.
—Gail Worcelo, from "Mary of the Cosmos"

A map of the world that does not include utopia is not worth even glancing at, for it leaves out the one country at which Humanity is always landing. And when Humanity lands there, it looks out, and, seeing a better country, sets sail. Progress is the realization of Utopias.
—Oscar Wilde

If by some miracle, and all our struggle, the Earth is spared, only justice to every living thing (and everything is alive) will save humankind.
—Alice Walker, quoted in *The Impossible Will Take a Little While*, ed. Paul Rogat Loeb

We cannot keep the birds of sadness from flying over our head, but we need not let them build nests in our hair.
—Chinese proverb

Our brains function as holographic information processors. If this is true for us as individuals, then it makes sense that our collective mind and consciousness may work this way as well. Today more than six billion humans (and minds) inhabit our planet. . . . The minimum number of people required to "jump-start" a change in consciousness is 1% of the population. . . . Knowing that everything from the most horrible suffering to the most joyous ecstasy—and all the possibilities in between—already exist, we find that it makes perfect sense to bring those possibilities into our lives. And we do . . . through the silent language of imagination, dreams and belief.
—Gregg Braden, *The Divine Matrix*

Hope is believing in spite of the evidence, then watching the evidence change.
—Jim Wallis, *Faith Work*

We are not lacking in the dynamic forces needed to create the future. We live immersed in a sea of energy beyond all comprehension. But this energy, in an ultimate sense, is ours not by domination but by invocation.
—Thomas Berry, *The Great Work*

There is a movement to revere the awful centrifugal force of alienation, brokenness, division, hostility, and disharmony. God has set in motion a centripetal process, a moving toward harmony, goodness, peace and justice, a process that removes barriers.
—Desmund Tutu, *No Future Without Forgiveness*

Our goal should be to live life in radical amazement.
—Abraham Joshua Heschel

Great steps occur when the cosmic organizations go to another level of complexity. . . . These are what the French Jesuit and paleontologist Teilhard de Chardin called "creative unions." They bring into being something that never existed before. . . . Deep Reality is that place in the center of our being where we experience our existence in an unlimited way. . . . Every little thing counts because everything is real and is part of the picture. Nothing escapes; nothing is on the side. Everything is making its difference to the whole.

—Beatrice Bruteau, *God's Ecstasy*

Nearby is the country they call life. You will know it by its seriousness.
Give me your hand.

—Rumi

When our days become dreary with low-hovering clouds of despair, and when our nights become darker than a thousand midnights, let us remember that there is a creative force in this universe, working to pull down the gigantic mountains of evil, a power that is able to make a way out of no way and transform dark yesterdays into bright tomorrows. Let us realize the arc of the moral universe is long but it bends toward justice.

—Martin Luther King Jr., *Where Do We Go From Here: Chaos or Community?*

Hope . . . which whispered from Pandora's box only after all the other Plagues and sorrows had escaped, is the best and last of all things.

—Ian Caldwell and Dustin Thomason, *The Rule of Four*

The Light that flows through your system is Universal energy. It is the Light of the Universe. You give that Light form. What you feel, what you think, how you behave, what you value and how you live your life reflect the way you are shaping the light that is flowing through you. They are thought forms, the feeling forms and the outer forms that you have given to the Light. They reflect the configuration of your personality, your space-time being.

—Gary Zukav, *The Seat of the Soul*

The salvation of this human world lies nowhere else than in the human heart, in the human power to reflect, in human meekness and human responsibility. Without a global resolution in human consciousness nothing will change for the better and the catastrophe towards which this world is headed will be unavoidable.

—Václav Havel

We bang and bang on the door of hope, and don't anyone dare suggest there's nobody home.
—Barbara Kingsolver, *Small Wonders*

Every time people say yes to life in whatever form—the unborn life, life on death row, the life of the severely handicapped, the life of the broken and the homeless—they start to give hope to each other. . . . This hope can form a very strong bond among people who are willing to go where life is fragile and hidden.
—Henri Nouwen, *The Road to Peace: Writings on Peace and Justice*

One can only hope that our uniquely human qualities of adaptiveness, ingenuity, and opportunism will carry us through an uncertain and challenging future.
—Brian Fagan, *The Great Warming*

Hope is not the conviction that something will turn out well, but the certainty that something makes sense regardless of how it turns out.
—Václav Havel

Hope is one of our duties . . . part of our obligation to our own being and to our descendants.
—Wendell Berry

Deep down we know that the world is not really chaotic and capricious. We know that it has a rhythm to it even if it is one that we cannot hum or write down. We can grasp it, but not name it.
—Mark Ward, *Beyond Chaos*

Optimism is the belief that things will turn out well; but sad to say, the objective facts give little reason to expect that humanity will avoid environmental suicide. Hope, on the other hand, is an active, determined conviction that is rooted in the spirit, chosen by the heart, and guided by the mind. . . . Hope is the foundation of action.

—Mark Hertsgaard, quoted in *The Impossible Will Take a Little While*, ed. Paul Rogat Loeb

Hope, unlike optimism, is rooted in unalloyed reality. . . . Hope is the elevating feeling we experience when we see—in the individual's eye—a path to a better future. Hope acknowledges the significant obstacles and deep pitfalls along that path. True hope has no room for delusion.

—Jerome Groopman, *The Anatomy of Hope*

The word "hope" does not necessarily mean feeling reassured that all is well. Hope is more like figuring out, from a very black pit, where the ladder to the far-away light at the top is. It's coming to a realistic understanding of how things really work, so that we finally know what we're dealing with, and therefore where to begin.
—David Suzuki and Holly Dressel, *Good News for a Change*

When our ecological unconscious does not find its way to guide us as we grow and develop, it can contribute to the causes of suffering from failed relationships, depression, and fear of the natural environment rather than an affinity for it. Our emotional bond with the Earth is what we need to heal if we are to heal our planet and continue the work of unfolding the universe as God intended. . . . Our health can be seen as an indicator of the health of the planet. As we are depressed, so are the planet and all its living organisms and systems.
—Heather Leavitt

8

Sixteen Closing Prayers for Our Planet

Our hopes and our dreams go before us,
Our longing to heal, to be whole,
Together we birth a new moment,
The Great Work of hands, heart and soul.
—Diane Forrest, "We Are One"

God of all the earth,
you have given us the heritage
of this good and fertile land;
grant that we may so respect and use it
that others may thank us
for what we leave to them.
—New Zealand Prayer Book

Be praised, my Lord, for our Sister, Mother Earth.
Who nourishes and governs us,
And produces various fruits with many-colored
flowers and herbs.
Praise and bless the Lord,
And give thanks and serve him with great humility.
—St. Francis of Assisi, Canticle *of* the Sun

Grandfather,
Sacred One,
Teach us love, compassion, and honor
that we may heal the earth
and heal each other.
—Ojibway people of Canada

Master of the Universe, grant me the ability to be alone:
May it be my custom to go outdoors each day, among the trees and grasses, among all
growing things, there to be alone and enter into prayer.
There may I express all that is in my heart, talking with You, to Whom I belong.
And may all grasses, trees and plants awake at my coming.
Send the power of their life into my prayer, making whole my heart and my speech
through the life and spirit of growing things.
—Rabbi Nachman

This is the Earth, healed again, growing green and blue. I want you to remember this exactly as it is, and then go and tell the people that if enough of us hold this image in their minds, we can heal the Earth and make it like it was a long time ago.
—Grandfather Rolling Thunder

How strong and good
and sure your earth smells,
and everything that grows there.
Bless us, our land,
and our people.
Bless our forests with mahogany,
wawa and cacao.
Bless our fields
with cassava and peanuts.
Be with us in our countries
and in all of Africa,
And in the whole world.
—Ashanti prayer, Ghana

The lands around my dwelling
Are more beautiful
From the day
When it is given to me to see
Faces I have never seen before.
All is more beautiful,
All is more beautiful,
And life is thankfulness.
These guests of mine
Make my house grand.
—Eskimo blessing

Deep peace of the running wave to you.
Deep peace of the flowing air to you.
Deep peace of the quiet earth to you.
Deep peace of the shining stars to you.
Deep peace of the infinite peace to you.
—Gaelic rune

May the temporal structures we build
From rock, stone, brick and mortar
Reflect the divine intent of our lives
To be built into temples of God's Holy spirit.
As we seek to shelter God's spirit,
May we also seek to shelter the needs among and
within us:

Ridding the air and soil of toxic fumes and waste;
Our apartments of toxic mold and spores;
Our language and behavior of toxic words and deeds;
So that we may breathe the fresh exhilaration
Of new life in God.
—Arnold Isidore Thomas, "The Quarry of God's Justice," quoted in *Worship Ways*, United Church of Christ

May we turn inwards and stumble upon our true roots
In the intertwining biology of this exquisite planet.
May nourishment and power pulse through these roots,
And fierce determination continue the billion-year
dance.
—John Seed

Your creation teems with bondaged creatures,
great valleys become dumps,
great oceans become dumped pollution,
fish wrapped in dumped oil,
fields at a loss for dumped chemicals.
So we pray for creation, that has become a dump,
and for all your people
who have been dumped,
and dumped upon.
Renew your passion for life,
Work your wonders for newness,
Speak your word and let us begin again.
In your powerful presence, we resolve to do our proper work.

—Walter Brueggemann, "We Are Not Self-Starters," quoted in *Awed to Heaven, Rooted in Earth*

Every creature, every plant, every rock and grain of
sand proclaims the glory of its Creator;
Worships through color, shape, scent and form,
A multi-sensory song of praise.
Creator God, may we join with the whole of your creation in praising you, our Creator, through the fragrance and melody for our lives. Amen!
—Reuel Norman O. Marigza, from An Earth Liturgy

May the blessing of the earth be on you,
Soft under your feet as you pass along the roads,
Soft under you as you lie out on it, tired at the end of day;
And may it rest easy over you when, at last, you lie out
under it.
May it rest so lightly over you
That your soul may be out from under it quickly;
Up and off and on its way to God.
And now may the Lord bless you, and bless you kindly.
Amen
—Scottish blessing

Apprehend God in all things,
for God is in all things.
Every single creature is full of God
and is a book about God.
Every creature is a word of God.
If I spent enough time with the tiniest creature—
even a caterpillar—
I would never have to prepare a sermon. So full of God
is every creature.
—Meister Eckhart

Fire of Love . . . purify my heart.
Burning Bush . . . consume me.
Living Waters . . . wash over me.
Deep Well . . . draw me to you.
Spirit of Life . . . enliven me.
Breath of All Breath . . . breathe me forth.
Ground of All Being . . . root me in you.
Womb of All Life . . . birth me anew.
—Green Mountain Sisters, *Earth Prayer Beads*

Recommended Further Reading

Ackerman, Frank. *Poisoned for Pennies: The Economics of Toxics and Precaution.* Washington, D.C.: Island Press, 2008.

AtKisson, Alan. *Believing Cassandra: An Optimist Looks at a Pessimist's World.* White River Junction, VT: Chelsea Green Publishing, 1999.

Boice, Judith L. *The Art of Daily Activism.* Oakland, CA: Wingbow Press, 1992.

Bonzo, J. Matthew, and Michael R. Stevens. *Wendell Berry and the Cultivation of Life: A Reader's Guide.* Grand Rapids, MI: Brazos Press, 2008.

Cohen-Kiener, Andrea. *Claiming Earth as Common Ground.* Woodstock. VT: Skylight Pathways, 2009.

Corbett, Julia. *Communicating Nature: How We Create and Understand Environmental Messages.* Washington, D.C.: Island Press, 2006.

Ehrlich, Paul R., and Anne H. Ehrlich. *The Dominant Animal: Human Evolution and the Environment.* Washington, D.C.: Island Press, 2008.

———. *One With Nineveh: Politics, Consumption, and the Human Future.* Washington, D.C.: Island Press, 2004.

Gilbert, Elizabeth. *Eat, Pray, Love: One Woman's Search for Everything Across Italy, India and Indonesia.* New York: Penguin, 2007.

Goodall, Jane. *Harvest for Hope: A Guide to Mindful Eating.* New York: Warner Books, 2005.

Gore, Al, and Bill McKibben. *American Earth: Environmental Writing Since Thoreau.* New York: Library of America, 2008

Gore, Al. *An Inconvenient Truth: The Planetary Emergency of Global Warming and What We Can Do About It.* Emmaus, PN: Rodale Books, 2006.

Gottlieb, Roger. *A Greener Faith: Religious Environmentalism and Our Planet's Future.* New York: Oxford University Press, 2006.

Grossman, Elizabeth. *High Tech Trash: Digital Devices, Hidden Toxics, and Human Health.* Washington, D.C.: Island Press, 2006.

Hartmann, Thom. *Threshold: The Crisis of Western Culture.* New York: Viking Press, 2009.

———. *The Last Hours of Ancient Sunlight.* New York: Crown/Three Rivers Press, 2004.

Hawken, Paul. *Blessed Unrest: How the Largest Social Movement in History Is Restoring Grace, Justice, and Beauty to the World.* New York: Penguin, 2008.

Hays, Edward. *Prayers for a Planetary Pilgrim: A Personal Manual for Prayer and Ritual.* Leavenworth, KS: Forest of Peace, 1989.

Hightower, Jane M. *Diagnosis: Mercury, Money, Politics, and Poison.* Washington, D.C.: Island Press, 2008.

Homer-Dixon, Thomas. T*he Upside of Down: Catastrophe, Creativity, and the Renewal of Civilization.* Washington, D.C.: Island Press, 2006.

Jeavons, John. *How to Grow More Vegetables: Than You Ever Thought Possible on Less Land Than You Can Imagine.* Berkeley, CA: Ten Speed Press, 1982.

Johnson, Huey D. *Green Plans: Blueprint for a Sustainable Earth.* Lincoln: University of Nebraska Press, 2008.

Kennedy, Donald, ed. *Science Magazine's State of the Planet*

2008–2009. Washington, D.C.: Island Press, 2008.

LaChapelle, Dolores. *Sacred Land, Sacred Sex, Rapture of the Deep: Concerning Deep Ecology and Celebrating Life.* Durango, CO: Kivaki, 1998.

Louv, Richard. *Last Child in the Woods: Saving Our Children from Nature-Deficit Disorder.* Chapel Hill, NC: Algonquin Books, 2005.

Macy, Joanna. *World as Lover, World as Self: Courage for Global Justice and Ecological Renewal.* Berkeley, CA: Parallax Press, 2007.

McDonough, William and Michael. *Cradle to Cradle: Remaking the Way We Make Things.* New York: North Point Press, 2002.

McKibben, Bill. *The Bill McKibben Reader: Pieces from an Active Life.* New York: Holt Paperbacks, 2008.

———. *Deep Economy: The Wealth of Communities and the Durable Future.* New York: Holt Paperback Books, 2008.

———. *The End of Nature.* New York: Random House Trade Paperbacks, 2006.

———. *Enough: Staying Human in an Engineered Age.* New York: Times Books, 2003.

———. *Hope, Human and Wild: True Stories of Living Lightly on the Earth.* Minneapolis: Milkweed Editions, 2007.

Nabban, Gary. *Where Our Food Comes From.* Washington, D.C.: Shearwater, Island Press, 2008.

Plant, Judith, ed. *Healing the Wounds: The Promise of Ecofeminism.* Gabriola Island, B.C., Canada: New Society Publishers, 2008.

Register, Richard. *Ecocities: Rebuilding Cities in Balance with Nature.* Gabriola Island, B.C., Canada: New Society Publishers, 2006.

Robert, Karl-Henrik. *The Natural Step: Seeding a Quiet Revolution.* Gabriola Island, B.C., Canada: New Society Publishers, 2008.

Sachs, Jeffrey, *Common Wealth: Economics for a Crowded Planet.* New York: Penguin, 2009.

Seed, John, Joanna Macy, Pat Fleming, and Arne Naess. *Thinking Like a Mountain: Towards a Council of All Beings.* Gabriola Island, B.C., Canada: New Society Publishers, 2008.

Starhawk. *Webs of Power: Notes from the Global Uprising.* Gabriola Island, B.C., Canada: New Society Publishers, 2008.

Steffen, Alex. *Worldchanging: A User's Guide for the 21st Century.* New York: Abrams, 2008.

Swimme, Brian. *The Universe Is a Green Dragon.* Santa Fe: Bear and Co., 1984.

Tucker, Mary Evelyn. *Worldly Wonder: Religions Enter Their Ecological Phase.* NY: Open Court, 2003.

Taylor, Sarah McFarland. *Green Sisters: A Spiritual Ecology.* Cambridge, MA: Harvard University Press, 2007.

Weisman, Alan. *Gaviotas: A Village to Reinvent the World.* White River Junction, VT: Chelsea Green Publishers, 2008.

Williams, Terry Tempest. *Finding Beauty in a Broken World.* New York: Pantheon, 2008.

Wilson, Edward O. *The Future of Life.* New York: Alfred A. Knopf, 2002.

Worldwatch Institute. State of the World, 2009.

Yudelson, Jerry. *Choosing Green.* Philadelphia: New Society Publishers, 2008.

———. *The Green Building Revolution.* Washington, D.C.: Island Press, 2008.

———. *Green Building Trends: Europe.* Washington, D.C.: Island Press, 2009.

Disclaimer

As in all collections of this nature, every attempt has been made to provide sources for unattributed quotes. Alas, it is not always possible, but we will be happy to fill in any omissions—provided authorship can be verified—in future printings. Thank you for your understanding.

About the Author

After a distinguished career in business book publishing, Karen Speerstra founded the publishing consulting business Sophia Serve, where she coaches and edits writers from around the country. She is also the author of several books and lives on a small mountain near Burlington, Vermont, where, in addition to writing and editing, she teaches and plays the piano and gardens and tends her nationally registered Chartres-style labyrinth.

To Our Readers

Conari Press, an imprint of Red Wheel/Weiser, publishes books on topics ranging from spirituality, personal growth, and relationships to women's issues, parenting, and social issues. Our mission is to publish quality books that will make a difference in people's lives—how we feel about ourselves and how we relate to one another. We value integrity, compassion, and receptivity, both in the books we publish and in the way we do business.

Our readers are our most important resource, and we value your input, suggestions, and ideas about what you would like to see published. Please feel free to contact us, to request our latest book catalog, or to be added to our mailing list.

Conari Press
An imprint of Red Wheel/Weiser, LLC
500 Third Street, Suite 230
San Francisco, CA 94107
www.redwheelweiser.com